# Meeting the Energy Challenge
# A White Paper on Nuclear Power
# January 2008

## Department for Business, Enterprise & Regulatory Reform

*Presented to Parliament by the Secretary of State for*
*Business, Enterprise & Regulatory Reform*
*By Command of Her Majesty*

*January 2008*

CM 7296                              £33.45

# Contents

# Foreword by the Prime Minister

 Climate Change is quite simply the biggest challenge facing humanity. The latest report from the Inter-Governmental Panel on Climate Change leaves little doubt that human activity, and in particular greenhouse-gas emissions, is changing the world's climate, with potentially devastating consequences. But we have choices and decisions to make about how we move towards a low-carbon economy.

I am determined that the Government will provide strong leadership in meeting not only the challenge of climate change, but in addressing the imperative of ensuring secure energy supplies. This means having reliable access to the energy we need to power our economy, at affordable prices.

To meet this challenge we need to take determined long-term action to reduce carbon emissions in every aspect of the way we live, the way we use energy and the way we produce energy, including the way we generate electricity. That is why the Government has today concluded that nuclear should have a role to play in the generation of electricity, alongside other low-carbon technologies. We have therefore decided that the electricity industry should, from now on be allowed to build and operate new nuclear power stations, subject to meeting the normal planning and regulatory requirements.

Nuclear power is a tried and tested technology. It has provided the UK with secure supplies of safe, low-carbon electricity for half a century. New nuclear power stations will be better designed and more efficient than those they will replace. More than ever before, nuclear power has a key role to play as part of the UK's energy mix. I am confident that nuclear power can and will make a real contribution to meeting our commitments to limit damaging climate change.

January 2008

# Foreword by the Rt. Hon. John Hutton MP

Energy is an essential part of modern life. We need secure, clean and sufficient supplies if we are to continue to function as a modern society. But we face two long-term challenges:

- Tackling climate change by reducing carbon dioxide emissions both in the UK and abroad
- Ensuring the security of our energy supplies

There is no single, simple solution to these challenges. That is why our White Paper on Energy, "Meeting the Energy Challenge," set out a wide range of measures which together will set this country on the right course to meet our objectives.

In May 2007 we launched a consultation to examine whether nuclear power could also play a role in meeting these long-term challenges, alongside other low-carbon forms of electricity generation. We set out our preliminary view that it is in the public interest to give energy companies the option of investing in new nuclear power stations.

The purpose of the consultation was to subject this preliminary view, and the evidence and arguments for it set out in our consultation document, to a thorough and searching public scrutiny.

I would like to thank everyone who took part in our consultation, for sharing their views and for making clear their commitment to tackling the twin challenges of climate change and security of energy supply.

We received 2700 separate written responses to the consultation. A further 1600 people participated in meetings and events up and down the country. We have been greatly impressed by the quality of the responses we received and the contributions made at those meetings.

Following the consultation we have concluded that, in summary, nuclear power is:

- Low-carbon – helping to minimise damaging climate change
- Affordable – nuclear is currently one of the cheapest low-carbon electricity generation technologies, so could help us deliver our goals cost effectively
- Dependable – a proven technology with modern reactors capable of producing electricity reliably
- Safe – backed up by a highly effective regulatory framework
- Capable of increasing diversity and reducing our dependence on any one technology or country for our energy or fuel supplies.

However, it is clear from responses to the consultation that there are also widespread concerns about nuclear power. These demonstrated that concerns do not arise from a lack of knowledge but are genuine concerns which need to be properly addressed. Significant points were raised in the consultation about:

- the need to combat climate change and ensure secure energy supplies
- the adequacy of protection in the areas of safety, environmental release of radioactivity and national security
- the management of radioactive waste and particularly the need to make progress towards a long-term solution
- the appropriateness of relying on energy companies for the construction, operation and decommissioning of nuclear power stations
- the risk that cost over-runs in construction, in waste management and decommissioning will undermine the economic case for nuclear and could lead to costs falling on Government
- the perception that investment in nuclear energy will "crowd out" investment in alternative technologies, particularly renewables
- the argument that the contribution nuclear energy makes to the UK's overall energy mix is currently quite small, calling into question the materiality of any contribution nuclear might make in the future to tackling climate change and ensuring security of energy supplies
- the belief that there are better alternatives to nuclear which would also enable us to achieve our energy goals and that there should be a greater focus on saving energy
- among those supporting nuclear power, a concern about what was perceived as a growing skills gap in the nuclear industry.

And while the consultation responses showed that there is considerable support for nuclear power, many were prepared to support nuclear only on the basis that these concerns are adequately addressed.

The Government has considered all these points. There are two elements in our response. First, any contribution to meeting our objectives by nuclear power must be seen as one part of the overall approach. Our White Paper, "Meeting the Energy Challenge," sets out a range of measures, including measures to save energy and to strengthen the Renewables Obligation to ensure that renewable electricity plays a full role in taking the UK towards a low-carbon energy economy. The UK is committed to delivering its fair share of the European Council commitment to produce 20% of its energy from renewable sources by 2020. The Prime Minister announced last November that we will launch a consultation this year on how we are to achieve our targets, and publish our full renewable energy strategy in spring 2009 once the EU directive implementing the 20% target has been agreed.

Second, we have examined the specific concerns raised in the consultation and the extent to which they can be met by the existing regulatory framework, or could be met through further development of our policies. Specifically we have taken the view that we should act to ensure that there:

- is a clear strategy and process for medium and long-term waste management, with confidence that progress will be made
- are new legislative provisions setting out a funding mechanism that requires operators of new nuclear power stations to make sufficient and secure financial provision to cover their full costs of decommissioning and their full share of costs of waste management, and

- is a further strengthening of the resources of the Nuclear Installations Inspectorate (NII) to enable it to meet a growing workload.

Having reviewed the evidence, and taking account of these points, the Government believes nuclear power should be able to play a part in the UK's future low-carbon economy. We have also carefully re-examined the impact of excluding nuclear power from our future energy mix. Our conclusion remains that not having nuclear as an option would increase the costs of delivering these goals and increase the risks of failing to meet our targets for reducing carbon dioxide emissions and enhancing energy security.

**The Government believes new nuclear power stations should have a role to play in this country's future energy mix alongside other low-carbon sources; that it would be in the public interest to allow energy companies the option of investing in new nuclear power stations; and that the Government should take active steps to facilitate this.**

These steps will include the Government taking forward regulatory processes and other steps, as follows:
- undertaking a Strategic Siting Assessment and Strategic Environmental Assessment
- meeting the requirements of European law that new nuclear practices should be required to demonstrate that their benefits outweigh any health detriments (the "Justification" process)
- ensuring that the regulators and particularly the NII are adequately equipped to review new build proposals through a process of Generic Design Assessment
- bringing forward legislation to ensure that the framework for funding decommissioning and waste management liabilities is clear and properly ensures that each nuclear operator meets its costs
- making use of the provisions of the Planning Bill to ensure that nuclear development projects are treated like other critical infrastructure projects and are dealt with effectively through the use of a National Policy Statement
- working to strengthen the EU Emissions Trading Scheme so that investors have confidence in a continuing carbon market when making decisions.

In addition, to give greater confidence to the public and to investors, we will work with the NII to explore ways of enhancing further the transparency and efficiency of the regulatory regime, without diminishing its effectiveness, in dealing with the challenges of new build.

This White Paper sets out the basis for our conclusion. It explains that we have a regulatory regime in the UK that can ensure that nuclear power remains safe and secure. We have made progress since 2003 towards a long-term solution to waste management. And we are confident that the new powers we are taking will ensure that industry will meet the full costs of decommissioning and their full share of waste management and disposal costs.

The Government has reached this decision in favour of allowing energy companies the option to invest in new nuclear power stations after careful thought and consideration of all the issues. Against the challenges of climate change and security of supply, I believe that the evidence in support of new nuclear power stations is compelling and that we should positively embrace the opportunity of delivering this important part of our energy policy.

January 2008

# The Structure of this White Paper

This White Paper sets out the decision we have taken in response to the consultation on nuclear power. It also examines the key concerns that emerged through the different strands of our consultation: we identify these in the analysis of responses to our consultation[1]. Further, it explains how we have addressed these issues in reaching our conclusion on nuclear power.

In Section 1 of this White Paper we summarise the consultation process which ran between May and October 2007[2]. This process is explained in more detail in our analysis of inputs. In Section 2 we address in detail the key issues which arose from our consultation, and explain how we have taken them into account in shaping our policy, and reaching our conclusions. In Section 3 we set out the facilitative actions that the Government will take, as we have done for other generation technologies[3], to reduce the regulatory and planning risks associated with investing in new nuclear power stations. Finally, there are three annexes:
- Annex A – Alternatives to Nuclear Power
- Annex B – Justification and Strategic Siting Assessment processes: this is the summary analysis of and formal response to the technical consultation which we ran concurrently with the main nuclear consultation
- Annex C – Regulatory and Advisory Structure for Nuclear Power.

Alongside this White Paper we are publishing on the BERR website:
- An analysis of consultation responses
- An Impact Assessment of our White Paper on Nuclear Power
- A flow diagram on UK energy supply and consumption which is relevant to the analysis contained in this White Paper of nuclear power and carbon dioxide emissions.

We have also published all written responses on our consultation website, except where individuals asked for their response to be treated as confidential[4].

1   The Future of Nuclear Power, *Analysis of consultation responses*, URN 08/534, January 2008.
2   The Future of Nuclear Power, *The Role of Nuclear Power in a Low Carbon UK Economy*, *Consultation Document*, URN 07/970, May 2007.
3   For more details of all our energy policies see the Energy White Paper, *Meeting the Energy Challenge*, URN 07/1006, May 2007.
4   www.direct.gov.uk/nuclearpower2007

**9**

# Overview

1.    Following our consultation on the future of nuclear power[5], the Government has reviewed the evidence and arguments referred to in the consultation document in the light of responses it received and in the light of any other evidence which has emerged. **The Government believes it is in the public interest that new nuclear power stations should have a role to play in this country's future energy mix alongside other low-carbon sources; that it would be in the public interest to allow energy companies the option of investing in new nuclear power stations; and that the Government should take active steps to open up the way to the construction of new nuclear power stations. It will be for energy companies to fund, develop and build new nuclear power stations in the UK, including meeting the full costs of decommissioning and their full share of waste management costs.** This White Paper[6] explains the basis for our decision, how we have considered responses to the consultation, and how we have taken them into account in framing our policy. We also explain in this White Paper what actions the Government will take to facilitate the construction of new nuclear power stations.

2.    This White Paper constitutes the Government's formal response to the nuclear consultation, and the related technical consultations on the proposed Justification and Strategic Siting Assessment processes. A full report on all responses to the consultation is available in the Government's analysis of consultation responses[7], which is being published simultaneously with this White Paper.

## The Government's energy strategy

3.    As explained in our consultation document, our decision and the steps we are taking will enable nuclear power to contribute to a low-carbon economy as part of our wider energy strategy.

4.    In 2006[8] the Government highlighted the challenges the UK faces in addressing climate change and ensuring security of energy supplies. In May 2007[9] we set out a programme of action and a new international and domestic energy strategy to meet these challenges and deliver our four energy policy goals:
   * to put ourselves on a path to cutting the UK's man-made emissions of carbon dioxide ($CO_2$) – the main contributor to global warming – by some 60%[10] by 2050, with real progress by 2020

---

5   The Future of Nuclear Power, *The Role of Nuclear Power in a Low Carbon UK Economy*, Consultation Document, URN 07/970, May 2007.

6   The Overview sets out the main conclusions and identifies specific concerns but it does not list them all. Sections 2 and 3 of the White Paper contain fuller details.

7   The Future of Nuclear Power, *Analysis of consultation responses*, URN 08/534, January 2008.

8   Energy Review, *The Energy Challenge*, July 2006.

9   Energy White Paper, *Meeting the Energy Challenge*, URN 07/1006, May 2007.

10  Compared to 1990. The Government is asking the new Committee on Climate Change to advise later this year on whether the target should be increased to 80%.

- to maintain the reliability of energy supplies
- to promote competitive markets in the UK and beyond, helping to raise the rate of sustainable economic growth and to improve our productivity
- to ensure that every home is adequately and affordably heated.

5.  The fundamental principle of our energy policy is that competitive energy markets, with independent regulation, are the most cost-effective and efficient way of generating, distributing and supplying energy. In those markets, investment decisions are best made by the private sector and independent market regulation is essential to ensure that the markets function properly and in accordance with our wider social and environmental objectives, particularly tackling climate change. That is why we have taken action, both at home and internationally, to create a framework of incentives, rules and regulations that encourage energy saving and investment in low- carbon technologies.

6.  We have strengthened our policy framework to underpin energy security and drive the reduction of $CO_2$ emissions through the proposals we set out in our White Paper last year[11], reinforced by the new approach to carbon budgeting set out in the Climate Change Bill[12]. Our commitment to carbon budgeting and ensuring an effective carbon price signal will help us to meet our contribution to the EU's target for reducing greenhouse gas emissions by 2020. We shall continue to seek to influence the wider international community, notably in getting consensus on a post-2012 agreement to reduce emissions of greenhouse gases. The Energy White Paper[13] also sets out the measures we are taking at home to enable us all to become more energy efficient and to increase the supply of energy from low-carbon sources. These measures include:
    - strengthening the EU Emissions Trading Scheme (EU ETS) to build investor confidence in the long-term future of the carbon market
    - strengthening of the Renewables Obligation, increasing the Obligation to up to 20% and introducing banding
    - running a competition for a demonstrator project for Carbon Capture and Storage (CCS)
    - lowering planning barriers to the installation of domestic microgeneration of electricity
    - making it easier to find information and advice on distributed generation
    - a trial of "smart" meters to record energy use and enable consumers to manage their demand
    - raising building standards and the energy efficiency standards of the appliances we use in our homes and other buildings.

7.  Furthermore, once agreement has been reached on each Member State's contribution to the EU 2020 renewable energy target[14], we will bring forward appropriate measures, beyond those set out in the Energy White Paper, to increase the share of renewable energy in our mix by

11  Energy White Paper, *Meeting the Energy Challenge*, URN 07/1006, May 2007.
12  http://www.defra.gov.uk/environment/climatechange/uk/legislation/index.htm
13  Energy White Paper, *Meeting the Energy Challenge*, URN 07/1006, May 2007.
14  Spring European Council conclusions, 8/9 March 2007.

2020. In the meantime, the measures and market framework set out in the Energy White Paper allow us to make significant progress on this important agenda and we will continue to take binding measures through the Energy Bill.

8. We set out the Government's preliminary view on nuclear power in our Energy White Paper[15]. This explained how nuclear power related to our overall energy strategy. In particular, we highlighted the uncertainties we face in the availability and costs of the UK's energy supplies over the coming decades. We also need to respond to the challenges of climate change. These uncertainties relate to: future fossil fuel and carbon prices; how quickly we can achieve energy efficiency savings and the therefore likely levels of energy demand; the speed, direction and future economics of development of the renewables sector; and the technical feasibility of and costs associated with applying carbon capture and storage technologies to electricity generation on a commercial scale.

9. It is our view that, given these uncertainties, our energy strategy should be based on diversity and flexibility in the energy mix and has accordingly developed policies which keep open the widest possible range of low-carbon generating options. These options would include renewables and the use of gas and coal with CCS, as well as nuclear. Unnecessarily ruling out one of these options would, in our view, increase the risk that we would be unable to meet our climate change and energy security objectives.

10. At the Spring European Council in March 2007, an EU energy action plan was agreed underpinned by a number of ambitious climate and energy targets for 2020[16]. These included unilateral targets[17] to reduce EU greenhouse gas emissions by 20%, rising to 30% in the context of a post-2012 international agreement; a target of 20% of the EU's energy to come from renewable sources; and a target to increase energy efficiency by 20%. The Council also underlined the central role of the EU ETS in meeting the target to reduce emissions. The Commission is expected to announce detailed proposals for meeting the renewables and emissions targets and on the future of the EU ETS in early 2008.

11. The Climate Change Bill is aimed at putting into legislation $CO_2$ reduction targets of 26-32% by 2020 and at least 60% by 2050. Cutting UK $CO_2$ emissions by 60% by 2050 will require extensive changes at all levels in the UK's energy system: in electricity generation and transmission; in energy storage and efficiency. To increase the likelihood that we will meet these targets, and meet them in the most cost-effective way, we need to make significant improvements in energy efficiency and develop a wide range of low-carbon and energy efficiency technologies including renewables and CCS.

---

15  Energy White Paper, *Meeting the Energy Challenge*, URN 07/1006, May 2007.
16  Spring European Council conclusions, 8/9 March 2007.
17  Compared to 1990.

12. Developments such as these will have implications for all non-renewable technologies in the UK. In a rapidly changing world, the ambitious reduction targets for 2050 and beyond, which may need to be tightened rather than relaxed in the future, place emphasis on the need to minimise the risk of failing to meet the target and reducing the costs of doing so through having as many options available as possible. Furthermore, an increasing role for electricity, and an expanded grid, for example in the context of de-carbonising and electrifying our transport or heating systems, could actually lead to an increasing demand for all forms of low-carbon electricity, including nuclear power. Hence, the Government's view is that none of these policy developments constitutes a reason to deny energy companies the option of investing in new nuclear power stations.

## Why decisions on nuclear power are needed now

13. As we explained in our consultation document[18], energy companies will need to build around 30-35 GW of new electricity generating capacity over the next two decades. They will have to make around two-thirds of this investment by 2020. So investment decisions made in the next few years will affect our electricity generation infrastructure for decades to come.

14. Of the 22 GW of capacity that is likely to close over the next two decades, just over a half is from carbon intensive fossil-fuel generation and about 10 GW is from nuclear power and therefore low-carbon. Companies' decisions on the type of power stations they invest in to replace this capacity will have significant implications for the level of future carbon dioxide emissions particularly beyond 2020. Currently, nuclear power provides approximately 19%[19] of our electricity generation and 7.5% of total UK energy supplies[20] and 3.5% of total UK energy use[21]. Without our existing nuclear power stations, UK total annual carbon dioxide emissions from all energy use would be 5-12% higher than they are today if gas or coal power stations had been built instead[22]. A saving of 5% in our $CO_2$ emissions[23] is, for illustrative purposes, about the same as taking a third of the UK's 32 million cars off the road. However, based on published lifetimes, most of the existing nuclear power stations are due to close in the next two decades. Although life extensions are possible, they are not guaranteed. This adds urgency to the need to take vigorous action on many fronts if we are to achieve a low-carbon energy mix and secure energy supplies.

18 The Future of Nuclear Power, *The Role of Nuclear Power in a Low Carbon UK Economy*, Consultation Document, URN 07/970, May 2007.

19 The May 2007 consultation document stated that nuclear power accounted for around 18% of electricity, based on the latest energy statistics available at that time. The most recent published data now available, in the Digest of United Kingdom Energy Statistics 2007, shows that in 2006 nuclear power accounted for 19% of the electricity generated in the UK.

20 This figure is the total amount of fuel used to generate electricity taken as part of total energy supplies. This issue is further discussed in Section 2.

21 See the simplified flow diagram of UK energy supply and consumption 2006 showing the role of nuclear at http://www.berr.gov.uk/files/file43008.pdf.

22 Sustainable Development Commission, *The Role of Nuclear Power in a Low Carbon Economy, Paper 2: Reducing $CO_2$ emissions – Nuclear and the Alternatives*, March 2006.

23 5% of our $CO_2$ emissions equals 29Mt $CO_2$.

15. It takes a long time to plan and build nuclear power stations. This means that new nuclear generation can make only a limited contribution before 2020. We will need other technologies (e.g. gas, renewables and coal) in this period. But we will need new capacity beyond 2020. To meet our 2050 $CO_2$ reduction target, our view is the answer lies in having a diverse and flexible energy mix and a framework which opens up, rather than closes down, low-$CO_2$ options.

16. Since the decision to keep open the question of nuclear power was taken in 2003[24] we have:
    - seen increasing evidence of climate change and wider international recognition of the need for global action
    - observed significant changes in the economics of nuclear power relative to other electricity generation technologies, driven by greater than expected increases in fossil fuel prices, and the introduction of a market price for $CO_2$ which requires investors to take account of the cost of $CO_2$ emissions in their investment decisions. Both factors increase the relative costs of fossil fuel electricity generation
    - developed the belief, based on scientific consensus and experience from abroad, that geological disposal will provide a technically possible means of disposing of radioactive waste
    - established the independent Committee on Radioactive Waste Management (CoRWM), whose main recommendations on the best means of managing existing higher activity radioactive waste were accepted by the Government
    - re-constituted CoRWM to provide scrutiny and advice on the implementation of waste management policy
    - established the Nuclear Decommissioning Authority (NDA) with expertise in waste management
    - consulted on a framework for implementing long-term waste disposal in a geological disposal facility through the Managing Radioactive Waste Safely consultation (MRWS)
    - seen a number of energy companies expressing a strong interest in investing in new nuclear power stations globally and in the UK.

## Main themes in the consultation

17. The nuclear consultation showed support for the Government's preliminary view, but it also revealed a number of important concerns. A majority of people agreed that nuclear was acceptable in principle, but wanted to be satisfied that their concerns were adequately addressed. Our accompanying analysis document details these issues[25]. People were concerned about a number of key issues. These include the need to combat climate change and ensure secure energy supplies, and the adequacy of protection in the areas of safety, environmental release of radioactivity and security. The management of radioactive waste, particularly the need to make progress towards a long-term solution was raised by many respondents. Others questioned the appropriateness of relying on energy companies for the construction,

---

24  For more details of all our energy policies see the Energy White Paper, *Meeting the Energy Challenge*, URN 07/1006, May 2007.
25  The Future of Nuclear Power, *Analysis of consultation responses*, URN 08/534, January 2008.

operation and decommissioning of nuclear power stations. Others were concerned about the risk that cost over-runs in construction, waste management and decommissioning will undermine the economic case for nuclear and could lead to costs falling to the Government. It was suggested that investment in nuclear will "crowd out" investment in alternative technologies, particularly renewables. It was argued that the contribution nuclear makes to the UK's overall energy mix is currently quite small, calling into question the materiality of any contribution nuclear might make in the future to tackling climate change and ensuring secure energy supplies. Some argued that there are better alternatives to nuclear power which would enable us to achieve our energy policy goals and the need for a greater focus on measures to save energy. Among those supporting nuclear power, there was a concern about what was perceived as a growing skills gap in the nuclear industry. The responses by the Scottish Executive and the Welsh Assembly to the consultation are covered in Section 2.

18.    We have considered these issues very carefully and this White Paper explains how we have taken them into account[26].

## Nuclear power – the issues

19.    In our consultation document[27] we first set out the context for our energy policy, as it relates to climate change and energy security. We said that in reaching our preliminary view, we had considered a number of issues relating to nuclear power, and that the consultation document set out the information and evidence that the Government had considered in reaching its preliminary view. We asked 18 specific questions designed to probe our assessment of the evidence relating to each of those issues. Following the consultation, we have reviewed the evidence and arguments referred to in the consultation document in the light of responses we received and in the light of any other evidence which has emerged. We summarise below our assessment of the inputs to the consultation[28] on each of the issues considered in the consultation document. We then set out our response and the basis for our conclusion that energy companies should be allowed the option of investing in new nuclear power stations. In reaching that conclusion we have taken account of the conclusions reached in relation to specific issues. However, we should emphasise that in reaching our decision we have considered the issues in the round and have given greater weight to some issues than others. Section 2 of this White Paper sets out our analysis of consultation inputs and our responses in greater detail.

26  The issues are addressed under each of the questions we asked in our consultation document.
27  The Future of Nuclear Power, *The Role of Nuclear Power in a Low Carbon UK Economy, Consultation Document*, URN 07/970, May 2007.
28  See The Future of Nuclear Power, *Analysis of consultation responses,* URN 08/534, January 2008 for more details.

## Climate change and energy security

20.  Climate change and energy security are the two greatest energy challenges we face. Tackling these twin challenges must be the focus of our energy policy. Climate change will have far reaching consequences for the UK and the rest of the world. The growing scientific consensus points to the need for urgent action to reduce carbon dioxide emissions. The Stern Review of the Economics of Climate Change is one of the many influential studies that highlight the economic costs of failing to tackle climate change[29].

21.  The future pattern of energy supply and demand points to a growing mismatch between the regions where energy is needed and those where natural resources are located. The UK has historically met most of its energy needs from domestic sources. In the past, we did this with coal and more recently with oil and gas from the North Sea. However, as production from the North Sea declines, we will become more reliant on supplies of oil and gas from regions which include less stable parts of the world, and at a time of rising demand and prices. At the same time, almost a third of our coal fired power stations are likely to close for a variety of reasons, including environmental legislation[30], and by 2023, based on their published lifetimes, all but one of our nuclear power stations will have closed.

22.  We believe there is a compelling case for action to meet these twin challenges. Our international and climate change strategy for meeting the challenges is built around four main elements:
     •  promoting open, competitive energy markets in the UK and abroad
     •  taking action to put a value on carbon dioxide emissions
     •  promoting investment to accelerate the deployment of low-carbon energy technologies
     •  putting in place policies to improve energy efficiency.

23.  We set out further details of our strategy in our Energy White Paper in May 2007[31]. However, we are clear that energy efficiency and renewable technologies on their own will not be enough to meet the twin challenges of climate change and energy security.

24.  Among those who took part in our consultation there was clear recognition and support for our strategy. There was also concern about the need for concerted international action on climate change. The Government fully appreciates the importance of an international response to climate change. Action in the UK alone will have a limited direct impact on global greenhouse gas emissions. It is therefore important that we use UK success in cost-effective delivery of ambitious targets as part of a concerted campaign to secure international action. The UK will, through the EU and bilaterally, use its influence to encourage the United States of America, China, India and others to engage actively in a global effort to reduce greenhouse gas

---

29  The Stern Review, *The Economics of Climate Change*, October 2006.
30  Directive 2001/80/EC of 23 October 2001 on the limitation of emissions of certain pollutants into the air from large combustion plants (O.J. L309/1, 27.11.2007).
31  Energy White Paper, *Meeting the Energy Challenge*, URN 07/1006, May 2007.

emissions. We also acknowledge concerns raised about our increasing reliance on imported fuel. We are confident that the measures set out in our Energy White Paper and this White Paper will ensure our future energy security.

## *Our conclusion*

**Without a clean, secure and sufficient supply of energy we would not be able to function as an economy or as a modern society. Climate change represents a significant risk to global ecosystems, the world economy and human populations. The scientific evidence is compelling that human activities are changing the world's climate. Nuclear power represents a low-carbon form of electricity generation. The majority of the UK's nuclear power stations are due to close over the next two decades. Over the same period, the UK will become increasingly reliant on imports of oil and gas, and at a time of rising global demand and prices, and when energy supplies are becoming more politicised. So in delivering the energy we need to support our economy and our society, we face two major challenges: climate change and energy security.**

**As the Government stated in its consultation document, the aim of Government should be to continue to raise living standards and the quality of life by growing our economy, while at the same time using every unit of energy as efficiently as possible. We also need to transform the way we produce the energy we need for light, heat and mobility. The Government has reviewed the arguments and evidence put forward, and continues to regard climate change and the security of energy supplies as critical challenges for the UK. They require significant and urgent action and a sustained strategy between now and 2050.**

## Nuclear power and carbon emissions

25. Analysing $CO_2$ emissions throughout the lifecycle of nuclear power stations, including the studies referred to in the consultation document and considering the reasons why there are differences between studies, has enabled the Government to be confident in confirming its preliminary view that nuclear power is a low-carbon form of electricity generation that can make a significant contribution to tackling climate change. Our estimates[32] of lifecycle $CO_2$ emissions from nuclear power are conservative, prudent and defensible. Ruling out nuclear as a low-carbon energy option would significantly increase the risk that the UK would fail to meet its $CO_2$ reduction targets because we would be placing greater reliance on fewer technologies, some of which have yet to be proven on a commercial scale.

26. Some respondents to the consultation expressed concerns that nuclear power can make only a small contribution to reducing our $CO_2$ emissions. It is not the Government's position that nuclear power alone

---

32 Life-cycle Assessment, *Vattenfall's Electricity in Sweden*, January 2005; OECD/ IAEA, *Uranium 2005: Resources, Production and Demand*, June 2006; and British Energy, Technical Report, *Environmental Product Declaration of Electricity from Torness Nuclear Power Station*, May 2005.

is the answer to meeting our emissions targets. Rather, vigorous action is required on a range of fronts, covering both supply and demand. Our analysis shows that excluding new nuclear power stations from the energy mix increases both the costs of meeting long-term emissions targets and the risks that we will not meet them. We estimate that existing nuclear power stations save between 5-12% of the UK's total $CO_2$ emissions. Nuclear power can and does make a material contribution to meeting targets. We conclude that it would not make sense to forego its potential for continuing to contribute in the future merely on the grounds that it cannot on its own completely solve the challenge of meeting emissions targets.

## *Our conclusion*

**After reviewing the arguments and evidence put forward, the Government is satisfied that, throughout their lifecycle, the $CO_2$ emissions from nuclear power stations are low. On reasonable assumptions, these emissions are about the same as those of wind generated electricity, and are significantly lower than emissions from fossil-fuelled generation. The Government therefore concludes that new nuclear power stations could make a material contribution to tackling climate change. However, it also believes that such a contribution needs to be part of a wider strategy to cut emissions.**

# Security of supply benefits

27. The Government believes that increasing the number of generating technologies available would increase the diversity and reliability of our electricity generating mix. Diversity of energy sources can help to reduce our dependence on gas as reserves fall in the North Sea and reduce the impact on the UK should prices for fossil fuels rise globally. Nuclear power is a proven and reliable form of electricity generation world-wide. It is therefore important in maintaining our energy supplies. Conversely, without nuclear power, the UK would depend on fewer technologies which could expose us to greater risks to the security of our energy supplies.

28. We acknowledge that uranium for new nuclear power stations needs to be imported but sources of uranium are diverse and secure. Currently nineteen countries produce uranium. For the most part, the UK obtains its uranium from Australia. While existing global uranium reserves are expected to last at least 85 years at current extraction rates, several responses to our consultation pointed out that there is inevitably some uncertainty over how long reserves of uranium will last, given both the uncertainty in the extent of future global deployment of nuclear power and the possible lack of commercial incentives to prove new reserves. Uncertainty over future fuel sources is not unique to uranium – for example at current production rates, global oil reserves are projected to last 40 years. However, we conclude that this uncertainty is not such as to undermine the significant contribution to energy security that arises from having diverse energy supplies, including nuclear power.

## *Our conclusion*

**Having reviewed the arguments and evidence put forward, the Government concludes that allowing energy companies the option of investing in new nuclear power stations would help the UK to maintain a diverse mix of electricity generating technologies with the flexibility to respond to future developments that we cannot yet envisage. Allowing energy companies the option of investing would therefore make an important contribution to the security of our energy supplies.**

# Economics of nuclear power

29. Based on a range of scenarios, we have concluded that nuclear power is likely to be cost-competitive with other sources of electricity in most scenarios and particularly where there is a price put on $CO_2$ emissions. Even on cautious assumptions, the cost of nuclear energy compares favourably with other low-carbon electricity sources, although, in due course, it will be for energy companies to make investment decisions based on their analysis of the economics.

30. A number of concerns were raised in the consultation about the prospect of cost overruns, and uncertainty over the cost of capital. We acknowledge that major capital projects entail financial risk. Whether nuclear provides sufficiently attractive returns given its financing characteristics is a matter that investors will determine. It is ultimately for energy companies to make a judgement about the economics of nuclear power. However, on the basis of our cost-benefit analysis, we think that nuclear power is likely to be an attractive economic proposition to them.

31. In the light of points made in the consultation, we have re-examined carefully the basis of the cost-benefit analysis which we published alongside the consultation document. We have reviewed discount rates, decommissioning and waste management costs and insurance rates. We have cross checked our analyses against the concerns raised by people contributing to the consultation. Having carried out this analysis, which we detail in Section 2, we feel confident in reasserting our view that the economics of nuclear remain attractive, both from the standpoint of the potential investor and of the wider economy as a whole.

32. The Government recognises the importance of a clear carbon price framework for all low-carbon technologies, including nuclear power. We believe this is best achieved through the EU and internationally. We will therefore continue to work to strengthen the EU ETS to build investor confidence in the existence of a long-term multilateral carbon price signal. We will also keep open the option of further measures to reinforce the operation of the EU ETS in the UK should this be necessary to provide greater certainty for investors.

## Our conclusion

**We have reviewed the arguments and evidence put forward, and based on the conservative analysis of the economics of nuclear power, the Government concludes that, under the most likely scenarios for gas and carbon prices, nuclear power would yield economic benefits to the UK in terms of reduced emissions of $CO_2$ and improved security of supply. It is for investors to determine whether the financing characteristics of nuclear power provide sufficiently attractive returns. However, on the basis of our cost-benefit analysis, we believe that nuclear power is likely to be an attractive economic proposition to them.**

**The Government is committed to working to strengthen the EU's Emissions Trading Scheme (EU ETS) and to building investor confidence in a long-term multilateral carbon price signal. We will keep open the option of introducing further measures to reinforce the operation of the EU ETS in the UK should this be necessary to provide greater certainty for investors.**

## The value of having low-carbon electricity generation: nuclear power and the alternatives

33. It is difficult to predict how energy supply and demand and the electricity generation mix will develop over the very long term. The factors which contribute to this uncertainty include:
    - growth in energy demand
    - the cost and availability of fossil fuels, and
    - the cost and availability of emerging low-carbon technologies.

34. The economic modelling we carried out for the Energy White Paper[33] and our consultation[34] indicates that if we excluded nuclear as an option, meeting our carbon dioxide emissions reduction targets would be more expensive. We also observed that, without new nuclear power to deliver a low-carbon economy by 2050, we would have to place even greater reliance on some technologies that are as yet unproven technically and commercially. Our preliminary view was therefore that giving energy companies the option of investing in new nuclear power stations lowers the costs and risk associated with achieving our energy goals of tackling climate change and ensuring energy security.

35. In the light of concerns raised in several inputs to our consultation, we have considered the argument that new nuclear capacity could harm the prospects for other low-carbon technologies. Our expectation is that if we are to meet our long-term targets for $CO_2$, this will mean that both nuclear and renewable technologies could have a significant share of the market, together with fossil fuel generation coupled with CCS (assuming that CCS proves to be a viable technology). We have also continued to examine the impact of not allowing energy companies the

---

33 Energy White Paper, *Meeting the Energy Challenge*, URN 07/1006, May 2007.
34 The Future of Nuclear Power, *The Role of Nuclear Power in a Low Carbon UK Economy, Consultation Document*, URN 07/970, May 2007.

option to invest in new nuclear power stations[35]. Our detailed analysis (see Annex A) of the implications of going forward without nuclear power as an option, brings us to a number of conclusions:

- Reliance on single solutions for electricity supply will not allow us to meet our goals. The more diverse the range of options, the better placed the UK will be to deal with the possible futures that could unfold
- All sectors of the economy will need to contribute in the effort to reduce $CO_2$ emissions
- Without nuclear power as an option, it would take a greater effort to reduce emissions through more costly options both within and outside of the electricity generation sector; and we will have to rely on generation technologies, some of which, such as CCS, are as yet unproven on a commercial scale and which together have a less diverse set of characteristics
- Large changes would be needed in the electricity system in terms of the scale of new capacity needed: the EU 2020 Renewables targets will mean rapid deployment of renewable technologies in the medium term and learning how to maintain security of supply with large penetrations of wind and other intermittent renewable technologies, most likely through considerable investment in backup capacity. The overall challenges of delivering secure electricity supplies, while making the transition to the low-carbon economy, would be magnified over the long-term in the absence of a dependable low-carbon technology such as nuclear power. This would be particularly significant should safe and reliable CCS for power generation not be proven or deployed on a significant scale at reasonable costs.

36. In our analysis at Annex A we comment on the positive contribution of energy efficiency measures to reducing demand and hence carbon dioxide emissions as well as recognising the value in having decentralised electricity generation. However, we do not believe that these alone will be sufficient to achieve energy security.


## Our conclusion

**Having reviewed the arguments and evidence put forward, the Government believes that giving energy companies the option to invest in new nuclear power stations reduces the costs and risks associated with tackling climate change and ensuring energy security. Nuclear power needs to be part of an overall approach to electricity generation. We will also take further steps to support renewables, Carbon Capture and Storage and Distributed Generation as outlined in the Energy White Paper and implemented through the Energy Bill.**

---

35  Annex A is a continuation of the analysis set out in Chapter 5 of our consultation document – The Future of Nuclear Power, *The Role of Nuclear Power in a Low Carbon UK Economy, Consultation Document*, URN 07/970, May 2007.

**21**

## Safety and security of nuclear power

37.   The safety and security of nuclear power is of paramount concern and we have an effective regulatory framework in place to ensure that these risks are effectively managed and minimised.

38.   The consultation process and the deliberative events showed public concern about safety and security. There are risks, but we consider these are very low and that our regulatory arrangements address those risks. We accept that safety and security in relation to nuclear materials must be paramount and that our regulatory arrangements must ensure that this remains the case in all circumstances. Having reviewed the arguments and evidence put forward in the consultation and responses to it, the Government is satisfied that new nuclear reactors can be managed as effectively as existing nuclear power stations. Indeed, the evidence is that new nuclear reactors are designed to be safer than those currently operating. Our regulatory arrangements are effective and proportionate, and we need to strive to ensure they remain so. In Section 2 of this White Paper we show how, in forming our conclusions, we have taken into account the concerns which have been raised about safety, security and health impacts, as well as threats from terrorism.

39.   To ensure that the UK's regulatory regime can deal effectively with new and existing nuclear facilities, we have authorised the Health and Safety Executive (HSE) to increase the salary levels of its nuclear inspectors to ensure that it can recruit staff of the necessary calibre. We will work with the independent regulators to build on these developments by exploring ways of enhancing further the transparency and efficiency of the regulatory regime, without diminishing its effectiveness, in dealing with the challenges of new nuclear power stations.

### Our conclusion

**Having reviewed the arguments and evidence put forward, and based on the advice of the independent regulators, and the advances in the designs of power stations that might be proposed by energy companies, the Government continues to believe that new nuclear power stations would pose very small risks to safety, security, health and proliferation. We also believe that the UK has an effective regulatory framework that ensures that these risks are minimised and sensibly managed by industry.**

## Transport of nuclear materials

40.   The transport of nuclear materials carries with it some small risks to safety and security. However, these risks are well understood and the UK can draw on several decades of experience in managing them effectively.

22

41. Concerns were raised during the consultation about the ability of containers used in transit to withstand accidents or about the possibility that material in transit could be accessed by terrorists. Regular safety evaluations by the International Atomic Energy Agency (IAEA) and the European Commission ensure that nuclear transport procedures continually evolve to reflect the latest technological and scientific best practice. The Government continues to believe, therefore, that the risks of transporting nuclear materials are very small and that there is an effective regulatory framework in place for managing and mitigating those risks.

### Our conclusion

**Having reviewed the arguments and evidence put forward, and given the safety record for the transport of nuclear materials and the strict safety and security regulatory framework in place, the Government believes that the risks of transporting nuclear materials are very small and there is an effective regulatory framework in place that ensures that these risks are minimised and sensibly managed by industry. The Government believes that this is not a reason not to allow energy companies to invest in new nuclear power stations.**

## Waste and decommissioning

42. In our consultation document, we set out the Government's preliminary conclusion on waste and decommissioning:

   "The Government believes that new waste could technically be disposed of in a geological repository and that this would be the best solution for managing waste from any new nuclear power stations. The Government considers that waste should be stored in safe and secure interim storage facilities prior to a geological repository becoming available. We consider that it would be desirable to dispose of both new and legacy waste in the same repository facilities and that this should be explored through the MRWS process"[36].

43. The importance of securing effective long-term management of nuclear waste was a recurring theme in the consultation. The Government accepts that progress towards this must be a priority. It is essential that we deal with the significant quantity of legacy waste from past nuclear activities. However, we recognise that it is also essential to ensure there are safe and robust arrangements for dealing with new waste and spent fuel.

44. Following the work carried out by CoRWM, the Government's policy is that geological disposal, coupled with safe and secure interim storage, is the way forward for managing legacy waste[37]. There should be an orderly and progressive approach to locating, developing and

---

36  The Future of Nuclear Power, *The Role of Nuclear Power in a Low Carbon UK Economy, Consultation Document*, URN 07/970, May 2007.

37  Higher activity waste which includes ILW, HLW and could include spent fuel.

commissioning a geological disposal facility. Government also accepted CoRWM's recommendation that the process should be staged so as to incorporate a series of decision points. This will allow the programme and progress to be kept under review, including on cost and value for money grounds. As we made clear in the Managing Radioactive Waste Safely (MRWS) consultation,[38] we are committed to making further progress on delivering a long-term waste management solution through the MRWS programme. The Government's view of geological disposal, in light of the outcome of that consultation[39], is set out in Box 1.

45.  In the consultation on the future of nuclear power, the Government set out its preliminary view that new waste could technically be disposed of in a geological disposal facility and that this would be the best way to manage waste from new nuclear power stations. The Government also set out its view that waste should be stored in safe and secure interim storage facilities prior to a geological disposal facility becoming available. The consultation document also stated that it would be technically possible and desirable to dispose of both new and legacy waste in the same repository facilities and that this should be explored through the MRWS process.

46.  The consultation provided some support for the Government's preliminary view that geological disposal would be the best solution for managing new build as well as legacy waste. But many people felt that we had made insufficient progress towards a permanent solution for existing waste. Their view was that there should be further progress before energy companies should be allowed to invest in new nuclear power stations.

47.  Having taken account of the inputs to the consultation, we continue to believe that geological disposal would provide a technically possible way of disposing of existing and new radioactive waste. We have also concluded that it would be technically possible and desirable to dispose of both new and legacy waste in the same geological disposal facilities and that this should be explored through the MRWS process.

48.  We are also satisfied that there are feasible mechanisms for identifying a suitable site for a geological disposal facility, through the MRWS programme. We recognise that it will be many years before a geological disposal facility could be completed. We are satisfied that interim storage will provide an extendable, safe and secure means to hold waste for as long as it takes to identify a site for, and to construct a geological disposal facility.

49.  We have considered carefully whether it is right to allow operators to build new nuclear power stations before a geological disposal facility to take the waste arising from them is constructed. In practice, this will be many years in the future, so waiting for the completion of a geological disposal facility would prevent nuclear power from contributing to the new electricity generating capacity that we will need over the next 20 years as existing power stations, nuclear and fossil fuelled, reach the

38  Managing Radioactive Waste Safely, *A Framework for Implementing Geological Disposal*, 25 June 2007.
39  Managing Radioactive Waste Safely, *A Framework for Implementing Geological Disposal*, 25 June 2007.

end of their lives. Given the ability of interim stores to hold waste safely and securely, if necessary, for very long periods, we are satisfied that it is reasonable to proceed with allowing energy companies the option of investing in new nuclear power stations in advance of a geological disposal facility being available. On this basis we believe that it is right to confirm our preliminary view on the handling of waste from new nuclear power stations.

50.    Box 1 sets out the Government's statement on the MRWS process and geological disposal.

## BOX 1: GOVERNMENT STATEMENT ON THE MRWS PROCESS AND GEOLOGICAL DISPOSAL

- In October 2006, the Government[40] accepted the recommendation of the independent Committee on Radioactive Waste Management (CoRWM) that geological disposal was the best available approach to the long-term management of the UK's higher activity radioactive waste.
- CoRWM's recommendations followed more than two and a half years' work assessing all of the available options on the basis of a wide programme of engagement with the expert community, stakeholder groups and the public.
- CoRWM also recommended that progress towards geological disposal should be coupled with a robust programme of safe and secure interim storage. Again the Government accepted the Committee's recommendation saying that:

  "The design of new stores will allow for a period of interim storage of at least 100 years to cover uncertainties associated with the implementation of a geological repository. The replacement of stores will be avoided wherever possible, but the NDA will ensure that its strategy allows for a safe and secure storage of the waste contained within them for a period of at least 100 years".

- Delivery of these commitments by the Government and the Nuclear Decommissioning Authority (and its agents) will be supported by research and development programmes. Where appropriate, international programmes and experience will be drawn on. It is clear that geological disposal is the internationally preferred option for the long-term management of higher activity radioactive waste. There has been extensive progress towards delivery of geological disposal solutions internationally in recent decades. Within the next one or two decades, overseas geological disposal facilities are likely to become operational for spent fuel, in addition to the facilities that already exist for Intermediate Level Waste (ILW) and Low Level Waste (LLW).
- The Government also said in its response to CoRWM that it would explore the concept of voluntarism and partnership arrangements in delivery of geological disposal of the UK's higher activity radioactive waste. We set out proposals for doing this, and asked for people's views on the issue more widely in the June 2007 consultation document "Managing Radioactive Waste Safely: a Framework of Implementing Geological Disposal".

40   *Response to the Report and Recommendations from the Committee on Radioactive Waste Management (CoRWM).*

- This consultation closed on 2 November 2007. An analysis and summary of the responses has been published[41]. Overall there was general agreement with the Government's proposals, including that of seeking a voluntarism and partnership approach, although many detailed points were made.
- Following on from CoRWM's recommendation (in relation to existing waste), international opinion and in line with the MRWS consultation, the Government continues to see geological disposal as the way forward for the long-term management of the UK's higher activity waste.

51. Having recently completed the MRWS consultation on a framework for implementing geological disposal and the principles of voluntarism and partnership, the Government is satisfied that nothing has emerged which casts doubt on the feasibility of a geological disposal facility for new and legacy wastes. Through the MRWS programme we have the strategy and direction to deliver safe solutions for the management of the UK's new and legacy higher activity wastes. We are satisfied that this provides a feasible mechanism for identifying a suitable site for a geological disposal facility.

52. As we have said, the Government is satisfied that waste can be stored safely and securely on an interim basis for as long as it takes to complete a geological disposal facility. The Nuclear Decommissioning Authority (NDA), the implementing body for a geological disposal facility, is continuing research and development on waste management. The NDA is also carrying out a UK-wide review of interim waste storage provision to ensure that the Government policy of robust interim storage can be implemented until a geological disposal facility is available. Section 3 of this White Paper sets out that operators of new nuclear power stations will be required to pay for and ensure that interim storage is available for waste until we expect a geological disposal facility to be in a position to accept waste from new nuclear power stations and beyond that date to provide adequate contingency.

53. As set out in the consultation on The Future of Nuclear Power[42], on 26 March 2007 the Government announced an update of its policy for low level waste (LLW) management[43]. Under the new policy, the NDA is now responsible for developing and maintaining a national strategy for handling LLW from nuclear sites and for ensuring continued provision of the waste management and disposal facilities required. The LLW strategy that the NDA develops will be reflected in its annual plans and strategy document in due course, and which will be subject to public consultation.

54. The Government will put in place a framework through the Energy Bill to ensure that energy companies set aside sufficient funds to cover their decommissioning costs and their full share of waste management costs in a secure way. Further detail is set out in Section 3 of this White Paper.

41  *Summary and Analysis of Responses to the Consultation on Managing Radioactive Waste Safely; A Strategy for Implementing Geological Disposal*, January 2008.
42  The Future of Nuclear Power, *The Role of Nuclear Power in a Low Carbon UK Economy, Consultation Document*, URN 07/970, May 2007.
43  *Policy for the Long Term Management of Solid Low Level Radioactive Waste in the United Kingdom*, 26 March 2007.

55. In our consultation, we set out the Government's preliminary view on the ethical issues around whether to create new nuclear waste:

"There are also important ethical issues to consider around whether to create new nuclear waste, including the ethical implications of not allowing nuclear power to play a role, and the risks of failing to meet long-term carbon emissions targets. The Government has taken a preliminary view that the balance of ethical considerations does not require ruling out the option of new nuclear power. However, we intend that these ethical issues should be considered through this consultation document and respondents are invited to give their views."

56. The consultation set out to consider the ethical issues around whether to create new nuclear waste. The consultation also considered the ethical implications of not allowing nuclear power to play a role, and the risks of failing to meet long-term carbon emissions targets. Whilst the Government accepts that creating new waste raises ethical issues, we also agree with those who believe that nuclear power provides significant benefits for future generations as a low-carbon form of electricity generation and one that secures our energy supplies. On balance, we believe that not taking action now on climate change, by allowing energy companies to invest in new nuclear power stations, raises more significant inter-generational challenges in terms of climate change related $CO_2$ and on-going security of energy supplies, than does the management of radioactive waste. Thus the Government concludes that the balance of ethical considerations does not warrant ruling out the option of new nuclear power stations.

## *Our conclusion*

Having reviewed the arguments and evidence put forward, the Government believes that it is technically possible to dispose of new higher-activity radioactive waste in a geological disposal facility and that this would be a viable solution and the right approach for managing waste from any new nuclear power stations. The Government considers that it would be technically possible and desirable to dispose of both new and legacy waste in the same geological disposal facilities and that this should be explored through the Managing Radioactive Waste Safely programme. The Government considers that waste can and should be stored in safe and secure interim storage facilities until a geological facility becomes available.

Our policy is that before development consents for new nuclear power stations are granted, the Government will need to be satisfied that effective arrangements exist or will exist to manage and dispose of the waste they will produce.

The Government also believes that the balance of ethical considerations does not rule out the option of new nuclear power stations.

# Nuclear power and the environment

57.   In our consultation document we examined the environmental impacts that arise at different stages of the nuclear life cycle covering landscape and construction; water use and thermal discharge; mining and milling of uranium ore; and preparation of fuel for nuclear power. We also acknowledged that the most significant environmental challenge of nuclear energy lay in the management of radioactive waste produced by nuclear power stations. Waste management is discussed above and in Section 2 of this White Paper.

58.   We recognise and appreciate the concerns raised about the potential for accidents and their environmental consequences and about the environmental impact of uranium mining. We also acknowledge concerns related to the proposed reforms to the planning system, which echo responses to the Planning White Paper that the changes might remove the rights of local people in decisions on nationally significant infrastructure projects. Points were also made in response to the consultation about the landscape impacts of nuclear power stations in comparison to fossil fuel power stations and wind farms. These also noted that the land take of an on-shore wind farm can be much greater than a nuclear power station. We remain satisfied that stringent regulation here and overseas (where uranium is mined) provides adequate environmental safeguards to assess and mitigate the impacts. The Government has also undertaken to safeguard the engagement and consultation with communities affected by planning proposals in the Planning Bill. We will also carry out a Strategic Environmental Assessment (SEA), as part of the Strategic Siting Assessment (SSA).

## *Our conclusion*

**Having reviewed the arguments and evidence put forward, the Government believes that (with the exception of the waste issue discussed above) the environmental impacts of new nuclear power stations would not be significantly different to those of other forms of electricity generation and that they are manageable, given the requirements in place in the UK and Europe to assess and mitigate the impacts. Therefore, the Government believes that environmental impacts do not provide a reason not to allow energy companies the option of investing in new nuclear power stations.**

**In confirming the Government's view that it is in the public interest to allow energy companies the option of investing in new nuclear power stations, we propose to undertake a Strategic Environmental Assessment as part of a Strategic Siting Assessment.**

# The supply of nuclear fuel

59.   The UK currently relies on imports of uranium (mostly from Australia) for its existing nuclear power stations, although the NDA does own around 51,000 tonnes of uranium, which could be converted into

uranium based fuel or could be combined with the UK's 86.5 tonnes of plutonium and used to make Mixed Oxide Fuel. A recent report[44], commissioned for the NDA, estimates that the UK stocks of uranium and plutonium could fuel up to three 1000-MW reactors for 60 years.

60. There continues to be a lot of focus on increases in the price of uranium and speculation that uranium resources may not be sufficient to meet growing world demand. Backed up by a number of authoritative reports[45] including one from the Inter-Governmental Panel on Climate Change[46], the evidence shows that sufficient fuel will be available to fuel a new programme of nuclear power stations constructed in the UK. Furthermore, since the price of nuclear fuel represents a much smaller part of the cost of electricity than for other technologies, even significant price increases would have only a limited effect on overall generating costs.

61. More generally, the Government's view is that the developers of new power stations will bear the risks around uranium price and availability and make a judgement about the economic impacts this has on their investment appraisal.

### Our conclusion

**Having reviewed the arguments and information put forward, and based on the significant evidence that there are sufficient high-grade uranium ores available to meet future global demand, and the relatively small impact that allowing energy companies to invest in new nuclear power stations in the UK would have on global demand for uranium, the Government believes that there should be sufficient reserves to fuel any new nuclear power stations constructed in the UK.**

## Supply chain and skills capacity

62. The supply of key components and skills is material in considering how new nuclear power stations might be built in the UK. The Government has acknowledged (in our consultation document and in this White Paper – see Section 2) that the supply of both skilled people and equipment will be constrained at times and that action is required, in particular, to retain skills and train a new workforce. This is not simply a UK or nuclear issue: similar constraints are seen worldwide across the energy industry. We accept that the situation is challenging but this

---

44 *Uranium and Plutonium*, Macro-Economic Study, June 2007.
45 House of Commons Trade and Industry Committee, *New nuclear? Examining the issues*, Fourth Report of Session 2005–06. Volume I. July 2006; Euratom Supply Agency, *Annual Report 2006*; World Energy Council, *Energy and Climate Change*, June 2007; International Energy Agency, *World Energy Outlook 2006*; Australian House of Representatives, Standing Committee on Industry and Resources, *Australia's Uranium — Greenhouse Friendly Fuel for an Energy Hungry World*, November 2006; IAEA/OECD, *Uranium 2005: Resources, Production and Demand*, June 2006; Massachusetts Institute of Technology, *The Future of Nuclear Power: an interdisciplinary MIT study*, 2003.
46 R.E.H. Sims, et al, 2007: Energy supply. In Climate Change 2007: Mitigation. Contribution of Working Group III to the Fourth Assessment Report of the Intergovernmental Panel on Climate Change [B. Metz, et al], Cambridge University Press, Cambridge, United Kingdom and New York, NY, USA.

will be the case however we chose to meet our future energy supplies. For nuclear build, we think that the situation is manageable.

63.  The UK's nuclear sector is developing a strategy that will enable it to meet its future demand for skills. We see evidence, worldwide, of industrial investment to supply a nuclear renaissance, although we accept that demand is likely to run ahead of supply, at least some of the time, and that the non-nuclear equipment suppliers need to increase their investment. Furthermore, new nuclear power stations have long lead times, giving clear market signals during which construction and operational skills can be developed and long-lead equipment ordered. For these reasons, we believe that both the Government and project developers should keep the situation under review, but that the challenges can be managed effectively and do not constitute a reason to deny energy companies the option of investing in new nuclear power stations.

## *Our conclusion*

**Having reviewed the arguments and evidence put forward, the Government believes that the energy sector, nuclear and otherwise, faces challenges in meeting its need for skilled workers and in the capacity of the manufacturing supply chain to support new construction. However, we believe that the situation is manageable and that building new nuclear power stations does not present a significantly greater challenge than the alternatives. Indeed, a nuclear renaissance, here and around the world, presents opportunities for companies to grow and for individuals to have rewarding careers. We conclude, therefore, that the skills and supply chain situation does not provide a reason to prevent energy companies from investing in new nuclear power stations.**

# Reprocessing of spent fuel

64.  Spent fuel created by nuclear power stations may either be disposed of or recycled, through a process called reprocessing, to separate out the useful uranium and plutonium. Reprocessing of spent fuel has a number of advantages in that it maximises the recovery of the energy from the fuel, can improve energy security by providing a source of fresh fuel, and reduces the amount of high level waste. But there are a number of disadvantages. Reprocessing creates separated plutonium (which requires long-term storage) and other waste streams (including regulated discharges) and requires the transport of spent fuel and other nuclear materials.

65.  Our view remains that in the absence of any proposals from industry, new nuclear power stations built in the UK should proceed on the basis that spent fuel will not be reprocessed. As a consequence, plans for waste management and financing should proceed on this basis. This ensures that before any new nuclear power stations are built, we have a clear path for the handling of the waste that will be produced, and are confident as to its technical and economic feasibility.

Department for Business, Enterprise and Regulatory Reform MEETING THE ENERGY CHALLENGE

## Our conclusion

**Having reviewed the arguments and evidence put forward, and in the absence of any proposals from industry, the Government has concluded that any new nuclear power stations that might be built in the UK should proceed on the basis that spent fuel will not be reprocessed and that plans for, and financing of, waste management should proceed on this basis.**

**We are not currently expecting any proposals to reprocess spent fuel from new nuclear power stations. Should such proposals come forward in the future, they would need to be considered on their merits at the time and the Government would expect to consult on them.**

## Our proposals on nuclear power

66.  The Government has concluded that new nuclear power stations should have a role to play in our future energy mix alongside other low-carbon sources of electricity; that it would be in the public interest to allow energy companies the option of investing in new nuclear power stations; and that the Government should take active steps to facilitate its deployment. It remains a central plank of the Government's energy policy that competitive energy markets, with independent regulation, are the most cost-effective and efficient way of generating, distributing and supplying energy, to meet the twin challenges of tackling climate change and ensuring energy security.

67.  In reaching our conclusion we have carefully considered the evidence and arguments set out in the consultation document and have considered the responses to the consultation and any other relevant evidence which has emerged. In particular, we have considered a range of issues including:
     * nuclear power and carbon emissions
     * security of supply impacts of nuclear power
     * the economics of nuclear power
     * the value of having low-carbon electricity generation options: nuclear power and the alternatives
     * the safety and security of nuclear power
     * transport of nuclear materials
     * waste and decommissioning
     * nuclear power and the environment
     * the supply of nuclear fuel
     * supply chain and skills implications
     * reprocessing of spent fuel.

68.  These issues are discussed elsewhere in Section 2 of this White Paper. Having considered the issues in the round, we continue to believe that we face two long-term challenges namely, tackling climate change by reducing carbon dioxide emissions both in the UK and abroad, and ensuring the security of our energy supplies.

69.   There is also considerable uncertainty about the future energy mix, in particular, the pace of climate change and the pressures this will create, and geopolitical developments. There are also uncertainties relating to future fossil fuel and carbon prices; the speed at which we can achieve greater energy efficiency and therefore likely levels of energy demand here and globally; the speed, direction and future economics of development in the renewable sector; and the technical feasibility and costs associated with applying carbon capture and storage technologies to electricity generation on a commercial scale.

70.   In view of the need to meet our twin energy challenges and given the uncertainties about the future energy mix, we believe that preventing energy companies from investing in new nuclear power stations would increase the risk of not achieving our long-term climate change and energy security goals, or achieving them at higher cost.

71.   However, we recognise that there are significant concerns about a number of issues associated with nuclear power. For example, the public are concerned about risks in relation to safety, security, proliferation, transport and the environment. Whilst these are understandable concerns we think that the risks associated with nuclear power are small and that the existing regulatory regime is such that those risks can be effectively managed.

72.   The public is also concerned about the management of radioactive waste. We recognise the importance of having a mechanism for the long-term management of radioactive waste. We are satisfied that it would be technically possible to dispose of new nuclear waste in a geological disposal facility and that such waste could be stored safely and securely until such time as the geological disposal facility is ready. We are exploring this mechanism through the MRWS process and believe it will provide a feasible mechanism for identifying a suitable site for construction of a geological disposal facility.

73.   We recognise that there are also other concerns including concerns about the supply of uranium, skills and about the environmental impact of nuclear power. Whilst we accept that these are important issues, we think these issues can be managed and we do not think they provide a reason for not allowing energy companies to invest in new nuclear power stations.

74.   Having considered the issues above and the other arguments and evidence raised in the consultation and in the responses to it, we have concluded that it would be in the public interest to allow energy companies the option of investing in new nuclear power stations.

75.   The next steps the Government will take to facilitate investment in new nuclear power stations are outlined in Section 3 of this White Paper.

## *Our conclusion*

**In the context of tackling climate change and ensuring energy security, the Government has concluded that it would be in the public interest to give energy companies the option of investing in new nuclear power stations.**

# Other considerations

76. The specific comments raised by those responding to the consultation were diverse. Many of the issues raised are already addressed at the appropriate points elsewhere in this White Paper and we have therefore only dealt with certain points below.

77. There was no clear consensus about the need either to restrict new nuclear power stations to the vicinity of existing sites – though many respondents thought that this would be likely to happen naturally anyway – or to restrict them to approximately replacing existing capacity. On the latter point, the Government has therefore decided that no specific cap on future new nuclear capacity should be applied.

78. We expect that applications for building new power stations will focus on areas in the vicinity of existing nuclear facilities. Industry has indicated that these are the most viable sites. The suitability of sites will be assessed through the forthcoming SSA process. In addition to the SSA, any developer wishing to construct a new nuclear power station would also need to obtain relevant environmental, health and safety authorisations as well as development consent. We will consult on the criteria for assessing suitable sites and then on a draft list of sites. The Government will continue to monitor whether an appropriate market in suitable sites is developing.

79. We do not think it is appropriate to restrict new nuclear power stations to the replacement of existing capacity because the fundamental principle of our energy policy is that competitive energy markets, with independent regulation, are the most cost-effective and efficient way of generating, distributing and supplying energy. In those markets, investment decisions are best made by the private sector and independent regulation is essential to ensure that the markets function effectively.

80. We have also considered whether there is a need to impose any other restrictions on new nuclear power stations and have considered the comments made in response to the consultation. Many of the comments made have been addressed elsewhere in this White Paper and we do not address them all specifically here.

81. We have, however, considered whether it is necessary to take additional steps to promote investment in renewables, alongside nuclear. We have concluded that our plans to extend the Renewables Obligation level to 20%, subject to deployment, and to target additional support to help bring emerging technologies such as offshore wind and marine to market quicker, will adequately address this concern. We will

**33**

also bring forward further measures in the light of the EU's 20% renewables target for 2020.

82. We have also considered the concerns about ensuring that the private sector adequately provides for waste and decommissioning costs. This is why, in addition to the measures we will be taking in the Energy Bill, we have decided to create a Nuclear Liabilities Financing Assurance Board (NLFAB) as explained in Annex C[47].

## Our conclusion

**We are taking steps to facilitate nuclear new build as outlined in this White Paper. In addition, we are setting up the Nuclear Liabilities Financing Assurance Board (NLFAB), putting in place measures to ensure that the effectiveness of the Nuclear Installations Inspectorate is further enhanced, and reforming the planning system.**

**We think the Strategic Siting Assessment (SSA) and Strategic Environmental Assessment (SEA) processes will enable suitable sites to come forward. The Government will continue to monitor whether an appropriate market in suitable sites is developing. The Government expects that applications to build new nuclear power stations will focus on areas in the vicinity of existing nuclear facilities. However, we do not consider it is necessary to put in place additional restrictions or conditions before giving energy companies the option of investing in new nuclear power stations.**

## Opening up the way for new nuclear power stations

83. Many respondents supported the facilitative actions we proposed in our consultation document, including the proposals we set out in a separate consultation for Justification and a combined SSA and SEA.

84. We believe it is important to take action on a number of fronts to give confidence to investors by:
    - Strengthening the EU ETS so that investors have confidence in a continuing carbon price signal when making decisions
    - Improving the planning system for major electricity generating stations in England and Wales, including nuclear power stations, by ensuring it sets a framework for development consents that gives full weight to policy and regulatory issues that have already been subject to debate and consultation at a national level, and does not reopen these issues in relation to individual applications
    - Running an SSA process to develop criteria for determining the suitability of sites for new nuclear power stations and, combined with this, taking further the consideration of the high-level

47  We intend to create a new independent advisory body, the Nuclear Liabilities Financing Assurance Board to provide scrutiny and advice on the suitability of decommissioning programmes – see Box 4.

environmental impacts of new nuclear power stations through a formal SEA in accordance with the SEA Directive[48]

- Running a process of Justification (in accordance with the Justification of Practices Involving Ionising Radiation Regulations 2004)[49], to test whether the economic, social or other benefits of specific new nuclear power technologies outweigh the health detriments

- Assisting the nuclear regulators, to pursue a process of Generic Design Assessment (GDA)[50] of industry-preferred designs of nuclear power reactors to complement the existing site-specific licensing process

- Delivering legislative arrangements to ensure that operators meet their full decommissioning costs and their full share of waste management and disposal costs. This may also enhance investor confidence by giving greater certainty on how they will be expected to meet their liabilities.

## Implications for the market

85. If new nuclear power stations are to play a role in the future it is important that the market receives a clear signal now about whether or not it can be an investment option. This White Paper gives that signal. Our clear conclusion is that allowing energy companies to invest in new nuclear power stations will reduce the risk of not achieving our long-term goals on climate change and energy security, and will reduce the cost of doing so.

86. We have also established an indicative timetable showing the fastest practical route to the building of new nuclear power station (see paragraphs 3.7-3.9). We are confident that, by working with operators, whilst upholding the highest regulatory standards, we can deliver a framework that would enable energy companies to begin construction of the first new nuclear power station in the period 2013-2014.

87. For illustrative purposes, we set out overleaf in Chart 1 a potential path to new nuclear build, including our programme of facilitative actions.

48  Directive 2001/42/EC of 27 June 2001 on the assessment of the effects of certain plans and programmes on the environment (O.J. L197, 21.7.2001, p30).
49  Justification of Practices Involving Ionising Radiation Regulations 2004 (S.I. 2004/1769).
50  This is sometimes referred to generically as "pre-licensing".

## Chart 1: Indicative pathway to possible new nuclear power stations

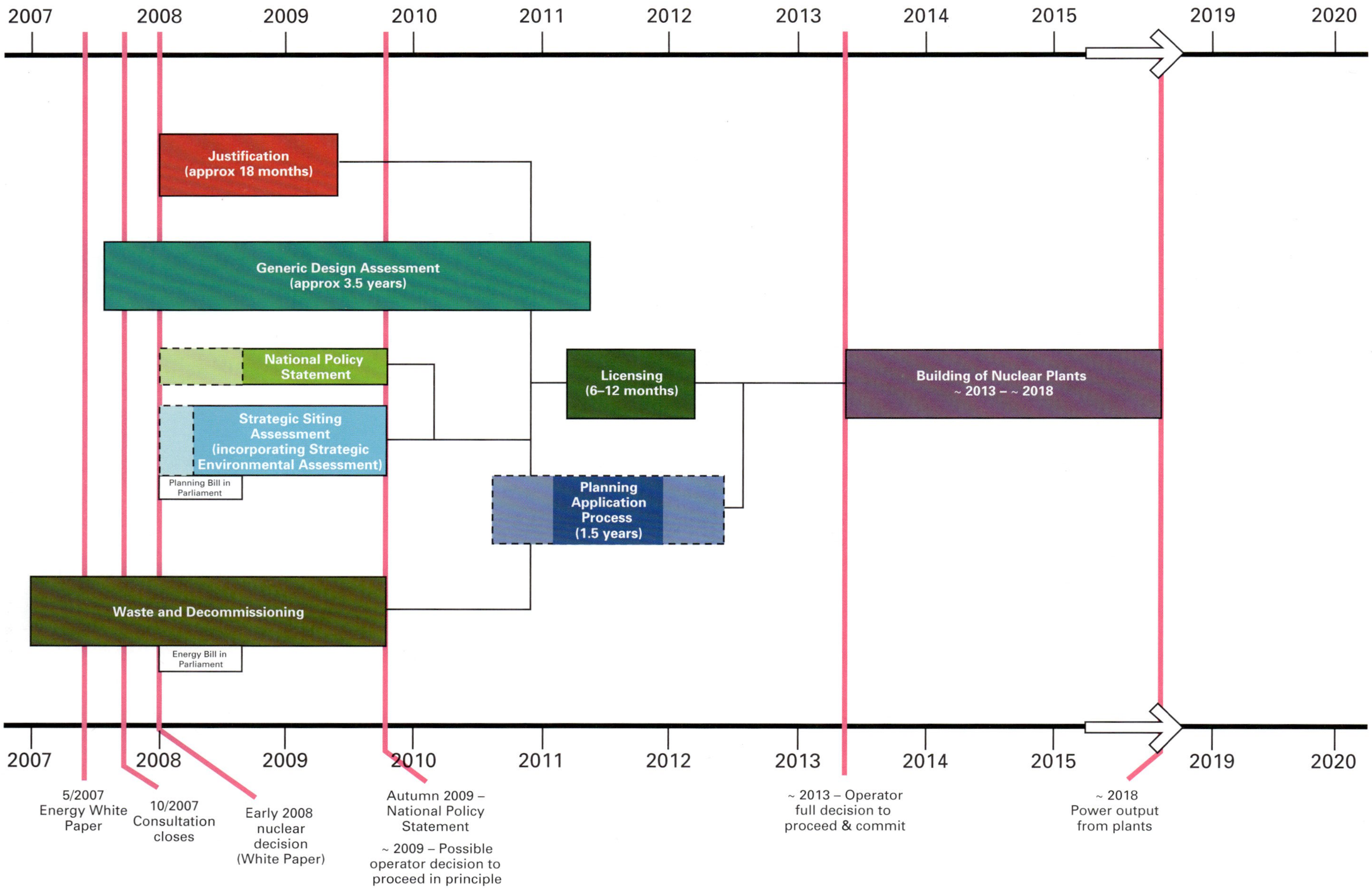

Timeline axis: 2007, 2008, 2009, 2010, 2011, 2012, 2013, 2014, 2015, 2019, 2020

**Justification** (approx 18 months)

**Generic Design Assessment** (approx 3.5 years)

**National Policy Statement**

**Strategic Siting Assessment** (incorporating Strategic Environmental Assessment)

Planning Bill in Parliament

**Licensing** (6–12 months)

**Planning Application Process** (1.5 years)

**Building of Nuclear Plants** ~ 2013 ~ 2018

**Waste and Decommissioning**

Energy Bill in Parliament

Power output from plants

~ 2018

~ 2013 – Operator full decision to proceed & commit

Autumn 2009 – National Policy Statement

~ 2009 – Possible operator decision to proceed in principle

Early 2008 nuclear decision (White Paper)

10/2007 Consultation closes

5/2007 Energy White Paper

36

# The Government's overall conclusion

88. **The Government has taken its decision to allow new nuclear power stations to be built against the very challenging backdrop of climate change and threats to our energy security. The Government's conclusion is that nuclear power is:**

   - **Low-carbon – helping to minimise damaging climate change**
   - **Affordable – nuclear is currently one of the cheapest low-carbon electricity generation technologies, so could help us deliver our goals cost effectively**
   - **Dependable – a proven technology with modern reactors capable of producing electricity reliably**
   - **Safe – backed up by a highly effective regulatory framework**
   - **Capable of increasing diversity and reducing our dependence on any one technology or country for our energy or fuel supplies.**

89. **On this basis, the Government believes it is in the public interest that new nuclear power stations should have a role to play in this country's future energy mix alongside other low-carbon sources; that it would be in the public interest to allow energy companies the option of investing in new nuclear power stations; and that the Government should take active steps to open up the way to the construction of new nuclear power stations. It will be for energy companies to fund, develop and build new nuclear power stations in the UK, including meeting the full costs of decommissioning and their full share of waste management costs. Together with the other policies set out in the Energy White Paper[51], a new nuclear programme will allow us to meet our wider energy goals. So that nuclear power can make the contribution of which it is capable, the Government will vigorously take forward the facilitative steps we describe in this White Paper.**

---

51 Energy White Paper, *Meeting the Energy Challenge,* URN 07/1006, May 2007.

# SECTION 1

# The Consultation Process

1.1    In 2003[52] the Government concluded:

> "Nuclear power is currently an important source of carbon-free electricity. However, its current economics make it an unattractive option for new, carbon-free generating capacity and there are also important issues of nuclear waste to be resolved. These issues include our legacy waste and continued waste arising from other sources. This white paper does not contain specific proposals for building new nuclear power stations. However, we do not rule out the possibility that at some point in the future new nuclear build might be necessary if we are to meet our carbon targets. Before any decision to proceed with the building of new nuclear power stations, there will need to be the fullest public consultation and the publication of a further white paper setting out our proposals"

1.2    This White Paper, and the public consultation which preceded it, are intended to fulfil the commitments made in 2003 that there would be the fullest public consultation before a decision was taken on the future of nuclear power. This section sets out the process which we followed to enable us to discharge the commitments we made in 2003 and how we set out to seek views on the information and arguments set out in our consultation document.

1.3    The detail of the consultation process is explained in more depth in the accompanying analysis document published alongside this document on the BERR web site[53]. That document analyses the inputs we received to the consultation and identifies the key themes which emerged from the consultation. That analysis underpins the Government's response to the consultation set out in this White Paper.

## Consultation on the future of nuclear power

1.4    The Government has carried out a series of consultations on the future of nuclear power. This process began prior to the Energy White Paper in 2003, continued with the consultation immediately before the publication of the Energy Review Report in 2006 and the further consultation after it was published, culminating in the consultation in May this year[54] setting out our preliminary view on nuclear power. This latest consultation, prompted by the ruling in the High Court in February 2007[55], takes account of that ruling and the Government's commitment in 2003[56] to the fullest public consultation and the publication of a further White Paper setting out our proposals for new nuclear power

---

52  Energy White Paper, *Our Energy Future – creating a low carbon economy*, February 2003.
53  The Future of Nuclear Power, *Analysis of consultation responses*, URN 08/534, January 2008.
54  The Future of Nuclear Power, *The Role of Nuclear Power in a Low Carbon UK Economy, Consultation Document*, URN 07/970, May 2007.
55  R (on the application of Greenpeace Ltd) v Secretary of State for Trade and Industry [2007] EWHC 311 (Admin).
56  Energy White Paper, *Our Energy Future – creating a low carbon economy*, February 2003.

stations in the event that a decision was taken in favour of nuclear power.

1.5 In the consultation document[57], we set out the Government's preliminary view that energy companies should have the option of investing in new nuclear power stations. The 2007 consultation set out the evidence that had been considered in reaching a preliminary view and considered, in the context of climate change and ensuring energy security, a number of issues relating to nuclear power:
- nuclear power and carbon emissions
- security of supply impacts of nuclear power
- the economics of nuclear power
- the value of having low-carbon electricity generation options: nuclear power and the alternatives
- safety and security of nuclear power
- the transport of nuclear materials
- waste and decommissioning
- nuclear power and the environment
- the supply of nuclear fuel
- supply chain and skills implications
- reprocessing of spent fuel.

1.6 The purpose of the May 2007 consultation was to ensure that in reaching a final view the Government could do so with the benefit of:
- the evidence and arguments set out in the consultation document being subject to the most searching public scrutiny, enabling the Government to review its position in the event that any of the evidence and arguments were shown to be wrong or incomplete
- being clear about the concerns that members of the public and stakeholders have about nuclear power: in reaching a decision in response to the consultation we have considered those issues and we have also considered to what extent existing policy, or developments of that policy, can meet these concerns
- having had an opportunity to raise and consider any factors which were not weighed up in forming the Government's preliminary view.

1.7 We wanted to understand the reason for people's views on nuclear power. We said that, when considering responses to the consultation and in deciding whether to confirm our preliminary view, we would give greater consideration to the arguments and evidence than to simple expressions of support or opposition to new nuclear power stations[58]. The consultation was not a referendum on nuclear power and we have been careful not to treat it as one. The numbers of people who may have agreed or disagreed with a specific point provide some useful context, but as our focus has been on evidence and analysis of the arguments, ultimately the numbers of respondents were not the determinant of our in-principle decision.

57  The Future of Nuclear Power, *The Role of Nuclear Power in a Low Carbon UK Economy, Consultation Document*, URN 07/970, May 2007.
58  On the inside cover of The Future of Nuclear Power, *The Role of Nuclear Power in a Low Carbon UK Economy, Consultation Document*, URN 07/970, May 2007.

1.8   When we launched our nuclear consultation, we said it would provide us with information which would help us to take the decision whether or not to allow energy companies to build new nuclear power stations in this country.

1.9   The consultation had a number of different strands to enable a wide range of people to respond:
- We published a comprehensive consultation document "The Future of Nuclear Power – *The Role of Nuclear Power in a Low Carbon UK Economy*", and a separate satellite consultation document "Consultations on the proposed processes for Justification and Strategic Siting Assessment"
- We set up an interactive web site which enabled people to respond directly online. The website was designed to make the consultation as accessible as possible, by breaking down the material contained in the formal consultation document into sections that respondents clicked through with dialogue boxes to capture views on each of the consultation questions[59]. During the consultation period 3,756 people registered on the site, 2,043 made on-line submissions and a further 685 responded by e-mail or on paper. There were only two small campaign responses, relating to the nuclear power stations at Wylfa and Dungeness
- We held 13 stakeholder meetings across the UK, including England, Scotland, Wales and Northern Ireland with the aim of listening to views from a range of stakeholders, including local authorities, green groups, energy companies, consumer groups, trade unions, faith groups and academics
- We held nine meetings with those community groups who live near existing nuclear sites
- We held nine simultaneous one-day deliberative workshops in nine towns and cities across the UK involving nearly 1,000 members of the public. This enabled the Government to listen to a demographically representative sample of the UK population. The purpose of the events was to understand the views of the public after they had heard the key arguments in the consultation. Specifically, we used the events to ensure that we understood the issues in relation to nuclear power that members of the public were concerned about. In reaching our decision in response to the consultation we have considered those issues and we have also considered to what extent existing policy, or developments of that policy, could address those issues
- We organised a Ministerial roundtable with 20 key stakeholders
- We engaged with relevant organisations such as the Youth Parliament in Strathclyde, the Prospect trade union, the Energy Institute and a collection of faith groups in Carlisle.

1.10  In addition, to raise awareness and participation in the consultation, we:
- Ran an information advertising campaign in the national and regional press inviting members of the public to respond to the consultation
- Sent direct mail to 5,000 grassroots and community organisations

59  All material shown on the website was the same as that contained in the consultation document.

- Arranged for the Secretary of State (John Hutton) to take part in a live web-chat hosted on the No 10 web-site, covering energy policy, mainly on nuclear power, and wider departmental issues.

1.11 The consultation also gave rise to a number of items and programmes in the media in newspapers, radio and television, and also featured in a number of conferences.

1.12 Alongside the main nuclear consultation we consulted separately on the proposed processes for two facilitative actions, namely on the Justification process[60] and a combined Strategic Siting Assessment and Strategic Environmental Assessment.

1.13 In total, over 4,000 individuals and groups responded to the consultation or attended one of our events. In our view this represents a significant response. We were impressed by the high quality of responses and the obvious care and time respondents took in giving us their views. Taking all the strands of the consultation together, the Government is satisfied that we have met our goal of conducting the fullest public consultation on nuclear energy in line with the commitment we made in 2003[61]. We are very grateful to all those who participated in the consultation[62]. Taken together with the publication of the Energy Challenge in July 2006, the decision making process has been thorough and lengthy. It is time now to make our decision on the future of nuclear power in this country.

1.14 It is a source of regret that some environmental non-Governmental organisations (NGOs) withdrew from the consultation process, particularly as the Government sees environmental concerns as being of great importance in the energy policy decisions we face. Nonetheless, we are satisfied that the environmental arguments both for and against nuclear power were thoroughly explored during the consultation. We are aware of some of the key arguments that those NGOs have raised in relation to nuclear power, including the alternative economic modelling undertaken by Greenpeace[63], and have sought to address these arguments in our response.

1.15 Alongside this White Paper we are publishing all the responses received to this nuclear consultation[64] except where respondents requested confidentiality. We also publish a full analysis of those responses, which catalogues them, draws out the key points which they made and also records the main findings from the deliberative events and the events for regional stakeholders and nuclear site stakeholders which we held during the consultation.

60  Under the Justification of Practices Involving Ionising Radiation Regulation 2004 (S.I. 2004/1769).
61  Energy White Paper, *Our Energy Future – creating a low carbon economy*, February 2003.
62  Most of the responses submitted to the consultation can be found on line at www.direct.gov.uk/nuclearpower2007, except where those submitting information have asked for it not to be made publicly available. This website also includes reports on regional stakeholder meetings and the deliberative events held during the consultation.
63  S. Thomas, P. Bradford, A. Froggatt, D. Milborrow, *The Economics of Nuclear Power*, Greenpeace International, May 2007.
64  The Future of Nuclear Power, *Analysis of consultation responses*, URN 08/534, January 2008.

1.16 We have undertaken additional work to examine further the consequences for our energy mix, costs, security of supply and $CO_2$ emissions if there were no new nuclear power stations in the UK. We set this out in Annex A of this White Paper.

## Managing Radioactive Waste Safely (MRWS) programme

1.17 Between 25 June and 2 November 2007, the Government also conducted an MRWS public consultation[65]. This consultation specifically considered the proposed implementation framework for the geological disposal of the UK's higher activity radioactive waste, including the approach to selection of a site for an eventual geological disposal facility. The timing of the MRWS consultation and the nuclear consultation were designed to enable respondents to both consultations to consider the information presented in both documents before responding. It has also enabled the Government to consider responses to both consultations before deciding its policy on the future of nuclear power. We were able to assess whether any points made in the MRWS consultation had implications for the in-principle nuclear decision that is the subject of this White Paper. A number of issues of relevance to new nuclear power stations emerged from the MRWS consultation. We consider these under waste and decommissioning in the Overview and in Section 2 of this White Paper.

## Further consultations

1.18 The Government is committed to carrying out further consultations on a number of the facilitative actions described in Section 3, including consultations on any draft Justification decision and as part of the Strategic Siting Assessment process. There will also be consultations on issues in relation to the implementation of our proposals for waste management and decommissioning. These consultations build on, and should be seen in the context of, previous consultations. In addition to this, before any nuclear power station could be constructed, it will also need to go through the planning application processes. (See Section 3 for details of reforms to the planning system).

65 Managing Radioactive Waste Safely, *A Framework for Implementing Geological Disposal*, 25 June 2007.

SECTION 2

# Nuclear Power: An assessment of the evidence and arguments put forward in the Consultation

## Our analysis

2.1   We have structured our analysis in this White Paper to ensure that we took full account of all elements of the consultation – on-line and written responses; the deliberative events involving the public and regional stakeholder and other meetings held across the UK. Consistent with our objectives for the consultation described in Section 1, we looked for arguments and evidence – particularly new arguments, evidence and information. For each issue raised, we have looked at it on its merits to determine if new and significant new arguments or evidence were presented. In this White Paper we have, in addition to recording the issues raised, shown how we have taken them into account in shaping our policy and we have recorded our conclusion. Our aim has been to capture in this White Paper the key issues raised by consultees, and in particular we have tried to focus on new or compelling arguments rather than every single point that had been raised, and show how we have taken them into account. In addition to reflecting such points, the White Paper also sets out what the Government believes to be the benefits of nuclear power.

2.2   Alongside this White Paper we are publishing a document[66] which analyses in greater detail the responses we received, on-line and on paper, alongside details of the deliberative events with the public and findings of the regional stakeholder events. We have also published all written responses on our consultation website, except where individuals asked for their response to be treated as confidential[67].

2.3   The remainder of this Section addresses each of the issues we have considered relating to nuclear power. In each case, we introduce each issue with a statement of our preliminary conclusions as put forward in our nuclear consultation document, and the accompanying question or questions (see shaded boxes). We also present the key arguments and issues that were presented in responses to the consultation, the Government's response, and our conclusion.

66  The Future of Nuclear Power, *Analysis of consultation responses*, URN 08/534, January 2008.
67  www.direct.gov.uk/nuclearpower2007

# Climate change and energy security

Energy is an essential part of everyday life in the UK. We use it to heat and light our homes, to power our businesses and to transport people and goods. Without a clean, secure and sufficient supply of energy we would not be able to function as an economy or a modern society. In delivering this energy we face two major challenges: climate change and energy security.

Climate change represents a significant risk to global ecosystems, the world economy and human populations. The scientific evidence is compelling that human activities, and in particular emissions of greenhouse gases such as carbon dioxide, are changing the world's climate. In 2005, 40% of global carbon dioxide emissions were created by the generation of electricity[68].

Temperatures and sea levels are rising. There is no scientific consensus on just how long we have to avoid dangerous and irreversible climate change, but the overwhelming majority of experts believe that climate change is already underway, and without action now to dramatically reduce carbon dioxide emissions, we will have a hugely damaging effect on our country, planet and way of life.

The Stern Review of the economic impacts of climate change[69] highlighted the need for an urgent, coordinated international response. The analysis is stark. It suggests that working together to mitigate the problems of climate change now would cost about 1% of global GDP per annum by 2050 with a range of +/-3% to take account of a number of variables including the availability of technologies. But as a comparison, it could cost around 5% of global GDP per annum in the long term if we do nothing. This cost could rise, to as much as 20% of GDP, if we take into account a wider range of issues such as human health and the environment.

Historically, the UK has met most of its energy needs from domestic sources: coal, until the middle of the 20th century, and since the 1970s, oil and gas from the North Sea have driven our economy. Since the 1950s, nuclear power, fuelled by imported uranium, has generated a significant proportion of our electricity, reaching a peak of nearly 30% of electricity output in the 1990s. Over the past decade nuclear power met about one-fifth of our electricity needs. If we had built fossil fuelled power stations rather than nuclear power stations, the UK's total carbon emissions from all sectors might have been 5% to 12% higher in 2004[70].

68  International Energy Agency (IEA), *World Energy Outlook 2006*.
69  The Stern Review, *The Economics of Climate Change*, October 2006.
70  Sustainable Development Commission, *The Role of Nuclear Power in a Low Carbon Economy, Paper 2: Reducing CO$_2$ emissions – Nuclear and the Alternatives*, March 2006.

In the future, the UK will increasingly depend on imported oil and gas at a time of rising global demand and prices, and when energy supplies are becoming more politicised. At the same time, we know that over the next two decades or so almost one third of our coal and oil fired power stations are likely to close because of environmental legislation, and while nuclear operators may achieve life extensions at the existing UK plants, all but one of our nuclear power stations are due to have closed by 2023, based on their published lives. This will create new risks that need to be managed by our energy strategy.

Our aim should be to continue to raise living standards and the quality of life by growing our economy, while at the same time cutting waste and using every unit of energy as efficiently as possible. But based on existing strategies to reduce energy demand, the IEA predict global energy consumption is likely to grow by about 50% by 2030[71]. Therefore we will also need to transform the way we produce the energy we need for light, heat and mobility.

**Question 1**
**To what extent do you believe that tackling climate change and ensuring the security of energy supplies are critical challenges for the UK that require significant action in the near term and a sustained strategy between now and 2050?**

## Key arguments and issues presented in responses

2.4     A number of those responding on-line to this question gave their views on nuclear power in general as well as on the Government's question on climate change and security of supply. The analysis of responses set out here deals specifically with views on the need to ensure security of supply and tackle climate change. We have considered comments that relate specifically to wider issues elsewhere in this Section under the relevant question.

2.5     Most of the people who answered the question agreed that climate change and security of supply are critical challenges for the UK. They agreed that the UK needs significant and urgent action, a concerted strategy going beyond 2050, and strong leadership.

2.6     Some questioned whether climate change was due to human activity or were doubtful that any $CO_2$ reductions made by the UK would have much effect without concerted action involving countries such as the United States of America, China and India. However, a number of those who questioned whether UK $CO_2$ reductions in isolation would make a difference, still supported reducing $CO_2$ emissions, whether in the UK or globally, on environmental grounds.

2.7     There were concerns about the UK's possible increasing dependence upon imported fuel and the impact that this could have on energy security. Some questioned whether the Government's goal of continued

---

71  International Energy Agency (IEA), *World Energy Outlook 2006.*

economic growth was right. They felt that promoting continued economic development was associated with an ever increasing demand for energy, rather than focussing on reducing consumption and changing lifestyles. They felt that a better balance between energy use and conservation should be struck.

2.8 Although some respondents agreed that climate change and energy security were critical challenges, views differed on how we could meet them. These issues are considered later in this Section.

## Government response

2.9 The Government notes that there was a clear recognition and support amongst respondents about the need to tackle climate change and ensure security of energy supply. The Government also notes concerns that economic growth would inevitably lead to increased energy consumption. However, the Government believes that without a healthy economy, the UK would not be in such a strong position to play a leading role in helping develop the new, innovative low-carbon forms of electricity generation needed to tackle climate change globally. The UK economy has grown by around 50% since 1990, while energy consumption has increased by only 10% and carbon dioxide emissions have declined by more than 6% over this period. We have therefore shown that we can continue to grow while reducing our emissions of carbon dioxide. There was concern expressed about the need for concerted international action on climate change. The consultation document set out the reasons for the Government's preliminary view on this issue and referenced the Stern Review[72] of the economic impacts of climate change, which highlighted the need for an urgent, co-ordinated international response. The Government fully appreciates the need for such international action and that action in the UK alone will have little impact on global emissions, unless it is part of a concerted international campaign. The UK will, through the EU, use its influence to encourage the United States of America, China and India and others to actively engage in a global effort to lower $CO_2$ emissions. We also acknowledge concerns raised about the UK's increasing reliance on imports of oil and gas: we believe that the measures set out in our Energy White Paper[73] and this White Paper will ensure our future energy security.

72  The Stern Review, *The Economics of Climate Change*, October 2006.
73  Energy White Paper, *Meeting the Energy Challenge*, URN 07/1006, May 2007.

## *Our conclusion*

Without a clean, secure and sufficient supply of energy we would not be able to function as an economy or as a modern society. Climate change represents a significant risk to global ecosystems, the world economy and human populations. The scientific evidence is compelling that human activities are changing the world's climate. Nuclear power represents a low-carbon form of electricity generation. The majority of the UK's nuclear power stations are due to close over the next two decades. Over the same period, the UK will become increasingly reliant on imports of oil and gas, and at a time of rising global demand and prices, and when energy supplies are becoming more politicised. So in delivering the energy we need to support our economy and our society, we face two major challenges: climate change and energy security.

As the Government stated in its consultation document, the aim of Government should be to continue to raise living standards and the quality of life by growing our economy, while at the same time using every unit of energy as efficiently as possible. We also need to transform the way we produce the energy we need for light, heat and mobility. The Government has reviewed the arguments and evidence put forward, and continues to regard climate change and the security of energy supplies as critical challenges for the UK. They require significant and urgent action and a sustained strategy between now and 2050.

# Nuclear power and carbon emissions

## THE GOVERNMENT'S PRELIMINARY VIEW

The Government believes that, based on the significant evidence available, the lifecycle carbon emissions from nuclear power stations are about the same as wind generated electricity with significantly lower carbon emissions than fossil fuel fired generation. As an illustration, if our existing nuclear power stations were all replaced with fossil fuel fired power stations, our emissions would be between 8 and 16 MtC[74] (million tonnes of carbon) a year higher as a result (depending on the mix of gas and coal-fired power stations). This would be equivalent to 30-60% of the total carbon savings we project to achieve under our central scenario from all the measures we are bringing forward in the Energy White Paper. Therefore, the Government believes that new nuclear power stations could make a significant contribution to tackling climate change. We recognise that nuclear power alone cannot tackle climate change, but these figures show that it could make an important contribution as part of a balanced energy policy.

### Question 2
**Do you agree or disagree with the Government's views on carbon emissions from new nuclear power stations? What are your reasons? Are there any significant considerations that you believe are missing? If so, what are they?**

## *Key arguments and issues presented in responses*

2.10 The consultation document presented a range of estimated $CO_2$ emissions for the entire nuclear lifecycle (i.e. including $CO_2$ emitted during construction, operation and decommissioning of the power station, mining, and transport of fuel, and disposal of waste). These estimates range from 7-22 g/kWh of electricity generated. We based this on data from the Organisation for Economic Co-operation and Development (OECD) and IAEA. This compares with 385 g/kWh for lifecycle $CO_2$ emissions from a gas-fired electricity power station, and 755 g/kWh for lifecycle $CO_2$ emissions from a coal-fired electricity power station[75]. Some respondents felt this overestimated the $CO_2$ emissions from the nuclear lifecycle; others felt it was a significant underestimate.

2.11 Those who saw the figures presented in the consultation document as overestimating $CO_2$ emissions from nuclear power cited a number of industry studies[76] based on lifecycle analyses at individual nuclear power stations, which showed much lower emissions than we had

---

74 The range of 8-16 MtC is for the impact on emissions of our existing nuclear power stations. This was based on the stations being replaced by either new gas (8 MtC) or new coal (16 MtC). If the nuclear stations were replaced by existing coal stations then the impact would be higher at 20 MtC.

75 *Nuclear Energy & the Kyoto Protocol* (2002) OECD Nuclear Energy Agency; *Assessing the difference: Greenhouse Gas Emissions of Electricity Generation Chains* (2000) IAEA Bulletin.

76 See footnotes 75 and 77.

estimated. For example, a study[77] by Vattenfall, the Swedish nuclear plant operator, at its Forsmark plant, suggested 3.10 g/kWh $CO_2$ emissions across the nuclear lifecycle. A lifecycle study by British Energy at its Torness station suggested $CO_2$ emissions of 5.05 g/kWh[78]. A number of references were also made to a survey of existing $CO_2$ emissions from various electricity sources, including nuclear power, carried out by the Parliamentary Office of Science & Technology in October 2006[79]. This gave a figure for nuclear of 5 g/kWh $CO_2$ emissions, rising to 6.8 g/kWh if lower grade ores were required. It was queried more generally whether the figures in the consultation document took full account of the more fuel efficient nuclear reactors and fuel enrichment processes now available, and also the potential for new nuclear power stations to be decommissioned with lower energy inputs than for previous generations. (Note: the figures presented in the consultation document were based on analysis of current $CO_2$ emissions from the nuclear lifecycle, and did not make assumptions about future potential efficiencies).

2.12   Those who felt the range of 7-22 g/kWh $CO_2$ emissions to be an underestimate, cited two studies in particular. The first was a 2005 study by van Leeuwen and Smith[80] which we had considered in formulating the Government's preliminary view in our consultation. The van Leeuwen/Smith study claims that current lifecycle $CO_2$ emissions from nuclear power are within the range of 84-122 g/kWh. The van Leeuwen/Smith study argues that data used by OECD, IAEA and other sources does not take into account the full energy costs of uranium mining, construction of nuclear power stations and their final decommissioning and waste storage or disposal. Moreover, van Leeuwen/Smith claim that there will be a significant depletion of high quality uranium ores over the lifetime of any new nuclear power stations. They believe that this will increase the carbon intensity of the nuclear lifecycle to such an extent that $CO_2$ emissions from nuclear would eventually surpass that of fossil fuelled power. We explain later (see paragraph 2.17 onwards) why we believe this claim is not credible.

2.13   The second main study cited by those who believed the 7-22 g/kWh $CO_2$ emissions figures in the consultation document to be an under-estimate was a 2006 study carried out by the University of Sydney. Commissioned by the Australian Government, this study suggests a range of $CO_2$ emissions of between 10-130 g/kWh, with an average lifecycle $CO_2$ emission rate of 60 g/kWh[81]. The authors of this study make clear that the very wide range of low to high figures presented is due in large part to the breadth of different assumptions that can be made when looking at $CO_2$ emissions and nuclear power. In particular, they depend on the assumptions made about the grade of uranium

77   Life-cycle Assessment, *Vattenfall's Electricity in Sweden*, January 2005.
78   British Energy, Technical Report, *Environmental Product Declaration of Electricity from Torness Nuclear Power Station*, May 2005.
79    Parliamentary Office of Science and Technology, *Carbon Footprint of Electricity Generation*, October 2006.
80   J.W.S van Leeuwen and P.Smith, *Nuclear Power: The Energy Balance*, August 2005 – a more recent summary of their case is provided in the report *Secure Energy? Civil Nuclear Power, Security and Global Warming*, Oxford Research Group, March 2007.
81   *Life-cycle Energy Balance and Greenhouse Gas Emissions of Nuclear Energy in Australia*, ISA, University of Sydney, November 2006.

**49**

ore used – the lower the grade of ore, the more energy required to mine and extract it in sufficient quantities. They also depend on which methods are used to enrich the fuel. The Australian study assumes a 70/30% split between the more efficient centrifuge enrichment and the less efficient diffusion enrichment techniques. The authors do acknowledge that the centrifuge technique will progressively account for a greater share of fuel enrichment.

2.14 The consultation document asserted that the lifecycle emissions of $CO_2$ from nuclear power were similar to those from wind power, taking into account the energy required at each stage of the lifecycle. Some respondents questioned whether the figures cited by the Government under-estimated emissions from wind power. In doing so, they expressed doubt over whether the figures took full account of all the new infrastructure that potentially has to be constructed for wind power, including standby generators (given the intermittency of wind power) and new grid connection, as well construction of the wind turbines themselves. Other respondents argued that lifecycle $CO_2$ emissions from wind power (and from other renewable sources) would decrease as the technologies and construction techniques improve.

2.15 Some of the respondents who agreed with the Government's view on nuclear power and $CO_2$ emissions nonetheless argued that the $CO_2$ emissions saved by nuclear were negligible given that nuclear-generated electricity accounts for around 3 to 4% of the UK's total energy use. It was argued that the time, money and resources required for new nuclear power stations would be better invested in renewables, carbon capture and storage technology and in greater energy efficiency measures, which would deliver more carbon savings in the long-term.

2.16 Some respondents commented that nuclear-generated electricity had an important and potentially growing role to play in providing low-carbon transport options, in particular through providing low-carbon electricity to power electric trains, trams, and cars. Therefore, in the future, the share of electricity in total UK energy use might increase.

## Government response

2.17 The Government recognises that some studies, mainly from industry, suggest lower $CO_2$ emissions for the nuclear lifecycle than the figures presented in the consultation document. At the same time a small minority of other findings – the van Leeuwen/Smith study in particular – present a comparably much higher estimate of $CO_2$ emissions across the nuclear lifecycle.

2.18 We continue to believe that the range of figures presented in the consultation document – 7-22 $gCO_2$/kWh emissions – represents a prudent and conservative judgement which is fully in line with authoritative research published by the OECD and the IAEA[82]. It is also broadly in line with analysis carried out by the World Energy Council

82 OECD *ibid.*, IAEA *ibid.*

(WEC) in 2004 which presented a range of $CO_2$ emissions of 5-40g kWh emissions for the nuclear lifecycle (with the highest figure being based on enrichment carried out entirely through the inefficient and increasingly obsolete diffusion method rather than the centrifuge process which is deployed at Capenhurst in the UK)[83]. The WEC's figures were endorsed by the Inter-Governmental Panel on Climate Change (IPCC)'s Working Group III in its Fourth Assessment Report of October 2007, which states that "Total lifecycle GHG (Green-House Gas) emissions are below 40 $gCO_2$/kWh (10 gC-eq/kWh), similar to those for renewable energy sources." The IPCC report goes on to say that "Nuclear power is therefore an effective GHG mitigation option"[84].

2.19 Lifecycle $CO_2$ emissions from nuclear power can vary depending on the assumptions used and methodologies employed. Unlike some industry and other studies, the figures in our consultation document assume that fossil fuelled power stations would continue to provide a significant proportion of the electricity used during parts of the nuclear lifecycle. The Government believes it is prudent to take account of this in estimating the range of potential $CO_2$ emissions across the lifecycle. Our estimates recognise that the emissions from fossil fuelled power stations are mostly at the point of generation whereas for nuclear power they occur during the mining and processing of uranium and during the construction and decommissioning of nuclear plants.

2.20 The van Leeuwen/Smith study has itself been subject to detailed critique in a range of studies[85]. Critics argue that it greatly overestimates the energy costs of mining lower grade uranium ores and also significantly overestimates the energy costs of constructing and operating nuclear power stations because of the particular methodology it employs. The study has additionally been criticised for assuming that there will be an inexorable decline in the availability of higher-grade uranium ore. As discussed elsewhere in this section in relation to the supply of fuel (paragraph 2.183), there is growing evidence that modern exploration techniques are identifying potentially significant new deposits of high-grade uranium ore. Additionally, new reactor designs and fuel enrichment techniques allow for much more efficient use of uranium ore than in the past. The IPCC Working Group III's Fourth Assessment Report of October 2007 looks at the future availability and utilisation of uranium resources and states that "Even if the nuclear industry expands significantly, sufficient fuel is available for centuries"[86].

2.21 We can confirm that the figures cited for wind power $CO_2$ emissions in the consultation document take account of the wind farm's full lifecycle, including intermittency rates which in turn determine back-up energy costs and emissions. The figures do not take into account the energy requirements for potential new grid connection but nor do the figures

83 *Comparison of Energy Systems using Life Cycle Assessment*, World Energy Council Special Report, July 2004.
84 R.E.H. Sims, et al, 2007: Energy supply. In Climate Change 2007: Mitigation. Contribution of Working Group III to the Fourth Assessment Report of the Intergovernmental Panel on Climate Change [B. Metz, et al], Cambridge University Press, Cambridge, United Kingdom and New York, NY, USA.
85 For example, see section 3.14 of *Life-cycle Energy Balance and Greenhouse Gas Emissions of Nuclear Energy in Australia*, ISA, University of Sydney, November 2006.
86 IPCC *ibid*.

**51**

for nuclear or other technologies presented in the consultation document.

2.22 With regard to those responses that queried how much of a role nuclear could make to reducing $CO_2$ emissions given its relatively low share of total energy use, nuclear power accounts for 19% of the electricity generated in the UK[87]. Electricity overall accounts for 18.5% of the energy used in the UK. Accordingly, nuclear generated electricity accounts for around 3.5% of total UK energy use (i.e. 19% of 18.5%) on a final consumption basis[88]. While the contribution that nuclear power currently makes to total energy consumption is relatively small, nuclear power has the potential in the future to reduce $CO_2$ emissions by more than its current share of the total energy mix. This is because the current electricity generation mix is very carbon intensive and nuclear power is one of very few proven low-carbon technologies in this sector and one that can provide low-carbon baseload electricity. Decarbonising this sector without new nuclear would be more difficult and more costly. We estimate that our current nuclear power stations save between 5-12% of the UK's total $CO_2$ emissions each year (assuming that the electricity would otherwise be generated by a mix of gas and coal-fired power stations), and analysis by the Sustainable Development Commission supports this[89].

2.23 This potential for nuclear to deliver significant $CO_2$ reductions well beyond its current proportionate share of the energy mix must also be seen in the context of the projected need to replace one-third of the UK's existing electricity generating capacity over the next two decades. Most of the existing nuclear power stations will close along with a number of oil and coal fired electricity power stations. Filling this major gap in electricity supply will be very challenging. Doing it in a way that meets the UK's $CO_2$ reduction goals will be even more challenging. It takes a long time to build new nuclear power stations. This means that new nuclear generation can only make a limited contribution before 2020. But we will need new low-carbon capacity beyond 2020 if we are to meet our 2050 $CO_2$ reduction target.

2.24 There is, however, no single answer. Our policy therefore is geared to tackling the issue in a number of ways, none of which would be sufficient on their own but which together represent a credible and deliverable response. Seen in this context, it would be illogical to rule out nuclear on the grounds that it makes only a modest contribution to our carbon emissions targets.

---

87 The May 2007 consultation document stated that nuclear power accounted for around 18% of electricity, based on the latest energy statistics available at that time. The most recent published data now available, in the Digest of United Kingdom Energy Statistics 2007, shows that in 2006 nuclear power accounted for 19% of the electricity generated in the UK.

88 The 3.5% figure is an energy output. In terms of energy inputs (i.e. the quantity of fuels used directly or expressed for comparative purposes, in terms of the equivalent amount of a standard fuel like oil), nuclear accounts for 7.5% of the UK's demand for primary fuels.

89 The Sustainable Development Commission identified a 5-12.6% saving in annual carbon emissions from nuclear power, depending on whether gas or coal power stations were used instead of nuclear (Sustainable Development Commission, *The Role of Nuclear Power in a Low Carbon Economy, Paper 2: Reducing $CO_2$ emissions – Nuclear and the Alternatives*, March 2006).

2.25 Ruling out nuclear power as a low-carbon option would significantly increase the risk of the UK failing to meet its long-term carbon reduction goals, in particular the target of reducing our $CO_2$ emissions by 60% by 2050. This is an ambitious and difficult goal and means that we need to maximise savings from all forms of low-carbon electricity generation, including nuclear power. We believe it would be short-sighted and wrong to exclude nuclear power – a proven means of abating $CO_2$ emissions which also delivers affordable electricity.

## *Our conclusion*

**After reviewing the arguments and evidence put forward, the Government is satisfied that, throughout their lifecycle, the $CO_2$ emissions from nuclear power stations are low. On reasonable assumptions, these emissions are about the same as those of wind generated electricity, and are significantly lower than emissions from fossil-fuelled generation. The Government therefore concludes that new nuclear power stations could make a material contribution to tackling climate change. However, it also believes that such a contribution needs to be part of a wider strategy to cut emissions.**

# Security of supply benefits

**THE GOVERNMENT'S PRELIMINARY VIEW**

The Government believes that the best way to achieve secure energy supplies is by encouraging a diversified mix of generating technologies, and that energy companies should have the widest choice of technologies in which to invest. We know that our nuclear power stations are coming to the end of their lives; not allowing energy companies to invest in new nuclear power stations would increase our dependence on fewer technologies and expose the UK to risks to the security of our energy supplies.

The Government believes that allowing energy companies the option of investing in nuclear power stations would make a contribution to maintaining a diverse generating mix, with the flexibility to respond to future developments that we cannot yet envisage. Allowing energy companies the option of investing would therefore make an important contribution to the security of our energy supplies.

**Question 3**
**Do you agree or disagree with the Government's views on the security of supply impact of new nuclear power stations? What are your reasons? Are there any significant considerations that you believe are missing? If so, what are they?**

## Key arguments and issues presented in responses

2.26 Those who supported the Government's view on security of supply felt that new nuclear power stations could make a positive contribution to maintaining a diverse energy mix in the UK. Some suggested that the most appropriate way to ensure energy security would be for demand to be low enough so that it can be met fully by renewable energy sources. Others offered the view that security could be better achieved by different renewable sources as well as new generations of clean burn fossil fuel technologies, such as CCS. Some were of the view that the best way to bring about diversity would be to promote novel renewable energy sources. This could include, for example, geothermal energy from Iceland and concentrating solar energy from North Africa as well as indigenous projects such as a Severn Barrage and other projects for tidal power and wave energy.

2.27 Some respondents pointed out that uranium has to be imported and supplies are finite so nuclear power can only be a temporary answer. Countries with uranium resources could increase prices as the global reliance on nuclear increases. As a result, it made sense to keep oil, gas and coal as back up. However, others pointed out that current nuclear plants could be seen to be unreliable as a base-load generation technology, pointing for example to recent shut downs of a number of British Energy nuclear power stations.

Department for Business, Enterprise and Regulatory Reform MEETING THE ENERGY CHALLENGE

2.28 A number of respondents highlighted the Government's claim that the UK will increasingly depend on imported gas, some of it from less stable regions of the world. They argued that gas mainly provides heat[90]. Only around a quarter of the gas used in the UK produces electricity. They also argued that nuclear power (alongside fossil fuels with carbon capture and storage and some renewables) would reduce the contribution of wind and wave power and have little or no effect on the amount of gas imported on the basis that nuclear power is ineffective in helping to balance the electricity grid.

## Government response

2.29 In the nuclear consultation document we outlined the three elements of electricity security of supply, namely capacity, diversity and a reliable supply chain. We explained how nuclear power contributes to enhancing security by affecting each of these.

2.30 A key element to ensuring sufficient capacity is to address future energy demand. Promoting energy efficiency will reduce emissions of $CO_2$ as well as the overall costs of providing the energy we need. Nevertheless, reducing energy demand is hugely challenging in the context of a growing economy, and even when all the measures we set out in the Energy White Paper in May 2007 are implemented, future electricity demand will at best stay flat[91]. Therefore, irrespective of these demand-side measures, the Government believes that even if demand does not increase, we will still need new electricity generation capacity to replace power plants as they close. To achieve our long-term carbon emissions reductions these new power stations will need to be low-carbon.

2.31 These new capacity requirements cannot fully be met through renewable sources due to the different types of generation necessary to ensure a flexible, secure mix. In future we will in fact need to replace both baseload and more flexible types of generation. Some renewables resources, such as wind or marine power, are variable in nature, and therefore we cannot rely on them exclusively to provide secure electricity supplies at times of peak demand. They would also be costly; the analysis undertaken for the Energy Review Report shows that generation costs from these technologies are higher than those for nuclear power[92].

2.32 Clean generation from fossil fuels, enabled by CCS could also help to provide the capacity the country will need. However, CCS represents a major technological challenge, as no commercial scale power station using CCS technology has yet been developed anywhere in the world, although all of the key elements of the individual stages of the process

---

90 For example, see report from the Science Policy Research Unit (SPRU), Sussex University http://www.sussex.ac.uk/sussexenergygroup/documents/security_brief_webonly.pdf.
91 The Energy White Paper, *Updated Energy and Carbon Emissions Projections*, URN 07/947, May 2007.
92 DTI, *Impact of banding the Renewables Obligation – Costs of Electricity production*, URN 07/948, April 2007.

have been demonstrated[93]. Therefore, there is a risk that CCS might not become a feasible generation technology in the time scale expected[94].

2.33 Some novel energy sources, e.g. electricity from geothermal sources in Iceland, could in principle help to meet our capacity requirements whilst also improving our technology and supply source diversity. However, they face formidable problems of cost. In fact, the Government's current approach does not preclude such developments and is it proposed to structure the Renewables Obligation to provide more support to new and emerging technologies that are further from the market[95]. The Government also helped to secure agreement with other EU member states at the 2007 Spring European Council to a binding target of 20% of the EU's energy consumption to come from renewable sources by 2020. The Government is also investigating the feasibility of novel projects at home, such as tidal technology and a Severn Barrage[96]. We will need to consider all these novel technologies in terms of their cost-effectiveness and their potential contribution to our energy policy goals including security of supply.

2.34 We recognise that nuclear power's base-load characteristic means that it is inflexible and cannot really be used to meet peak demand periods. To that extent, it is at a disadvantage compared to some other forms of electricity generation, but its main advantage is that it is a low-cost and dependable low-carbon source of generation. We continue to believe that nuclear power could help meet the expected requirements for new generation capacity over the next 20 years.

2.35 Diversity of electricity supplies increases the resilience of the system as it reduces the risks of interruptions and the risks of sudden and large spikes in electricity prices, which can arise when the system is excessively dependent on a particular technology or fuel. In this regard, nuclear has an advantage compared to fossil fuel generation (where fuel prices may increase dramatically within a short space of time) and thus might help to safeguard against the risk of short and long-term electricity price increases. Whilst the uranium price could increase, uranium is currently sourced from a diverse range of countries and is traded in a global market thus ensuring competitive prices. Even if uranium prices were to increase substantially, fuel costs make up only a small proportion (around 10%) of overall plant running costs, with uranium ore accounting for approximately 1.5% of total generation costs compared with gas plant where fuel costs represent around 70% of running costs. Fossil fuel prices have been volatile and subject to more sudden increases. Increases in fossil fuel costs are also more rapidly translated into increases in generation costs and electricity prices because fuel prices represent a higher proportion of the total cost of generating electricity. Nuclear power can therefore play a role as a hedge against such input price volatility.

---

93  See Annex B of Energy Review, *The Energy Challenge*, July 2006.
94  Project Information Memorandum, *Competition for Carbon Dioxide Capture and Storage Demonstration Project*, 19 November 2007.
95  Although renewable energy imported into the UK would not qualify under the Renewables Obligation.
96  See GNN, *John Hutton calls for open minds on the future of the Severn Barrage*, 25 September 2007.

2.36 Diversity of electricity supplies also contribute to the flexibility necessary to be able to respond to future developments in the energy and electricity sectors that we cannot yet envisage. Allowing nuclear to be an investment option therefore brings the potential for added value in managing the risks to security of supply which could feed through into the price of electricity.

2.37 Reliability in the fuel supply chain is another key element in achieving secure energy supplies. Nuclear fuel supply is a stable and mature industry. While uranium resources are finite, we need to put this in perspective against fossil fuel supplies. Uranium deposits are set to last much longer than oil and gas reserves; based on the levels of global nuclear generation in 2004, the known available reserves of uranium that can be mined for less than $130/kg (approximately the uranium price in 2006) would last for the next 85 years. In contrast, the current ratio of proven oil reserves to global oil production is around 40 years and the equivalent number for gas is around 63 years[97]. Moreover, exploration of uranium has been minimal in recent years because few new nuclear power stations have been built. We believe that a new nuclear build programme in the UK and globally could catalyse further exploration and development. The IEA have concluded[98] that world uranium resources are more than adequate to supply the expected global expansion of nuclear power.

2.38 We also recognise that, with no significant indigenous source of uranium ore, we will have to import uranium fuel. However, uranium imports come from a range of countries that are not necessarily the same as those that supply other energy sources. Uranium is currently mined in 19 different countries and resources of economic interest have been identified in at least 25 other countries. This provides valuable diversity of supply. We believe therefore that including new nuclear power as an option for investment would spread the supply risks that could be associated with a particular fuel or region of the world.

2.39 In terms of reliability, we recognise that the Advanced Gas Cooled Reactor (AGR) nuclear stations operated by British Energy have had a relatively poor record of reliability in recent years, with boiler problems as well as issues specific to the reactor type. However, Sizewell B, the most recent nuclear power station built in the UK, has a strong reliability record which it shares with modern Pressurised Water Reactors (PWRs) around the world. Over an 18 month period to December 2007 Sizewell B achieved a 89% load factor including a run of almost 400 days at full load following a planned outage.

2.40 Furthermore, allowing energy companies to develop and invest in a broad portfolio of different electricity generation technologies would not only increase the diversity and reliability of our electricity mix, but also the diversity and reliability of our energy mix as a whole, for example by reducing our dependence on one particular fuel, such as

97 *BP Statistical Review of World Energy 2007.*
98 International Energy Agency (IEA), *World Energy Outlook 2006.*

A White Paper on Nuclear Power

gas[99]. In 2006, the proportion of gas the UK burnt to produce electricity was 30% of total gas demand (i.e. 28.1 billion cubic metres). Much of the projected increase in gas demand to 2020 is also likely to come from the electricity generation sector as gas increases its share in the generation mix as existing coal and nuclear power stations close[100]. In 2020 the proportion of gas used for electricity generation could increase to around 38%.

2.41   At a time of falling reserves in the North Sea, this underlines the Government's concern about the need to diversify our sources of supplies and minimise our exposure to future import risks. A lower level of gas consumption in the future, for example as a result of higher levels of nuclear power generation, could reduce the need for investment in import and storage facilities to maintain gas security of supply.

2.42   Therefore, the Government continues to believe that the best way to achieve secure energy supplies is to maintain a market-based framework where energy companies have the opportunity to invest in the widest choice of technologies and sources of supply. We believe that allowing companies the option of building new nuclear power stations would help the UK to maintain a diverse electricity generation mix, and energy system as a whole. Such an approach would give us the flexibility to respond to future developments, for example, in the global markets for oil and gas.

## *Our conclusion*

**Having reviewed the arguments and evidence put forward, the Government concludes that allowing energy companies the option of investing in new nuclear power stations would help the UK to maintain a diverse mix of electricity generating technologies with the flexibility to respond to future developments that we cannot yet envisage. Allowing energy companies the option of investing would therefore make an important contribution to the security of our energy supplies.**

99   This is based on the assumption, as outlined in the Nuclear Power Generation Cost Benefit Analysis, that the alternative generation that would be built instead of new nuclear power would be gas-fired power stations.
100 According to latest projections gas power stations could provide around 53% of electricity in 2020.

# Economics of nuclear power

## THE GOVERNMENT'S PRELIMINARY VIEW

Based on a conservative analysis of the economics of nuclear power, as outlined in the consultation document, the Government believes that nuclear power stations would yield economic benefits to the UK in terms of reduced carbon emissions and security of supply benefits under likely scenarios for gas and carbon prices. As an illustration, under central gas and nuclear cases, and with a future carbon price of €36/t$CO_2$, the net present value over 40 years of adding 10 GW of nuclear capacity would be of the order of £15 billion.

### Question 4
**Do you agree or disagree with the Government's views on the economics of new nuclear power stations? What are your reasons? Are there any significant considerations that you believe are missing? If so, what are they?**

## *Key arguments and issues presented in responses*

2.43    There was general support, with qualification, for the Government's view. Some respondents, however, said that nuclear generation costs are typically underestimated and the assumptions on discount rates and financing period for nuclear are too optimistic while the discount rate for appraising wind generation costs is too high[101]. A further point was raised that the only new nuclear power station being built (in Finland)[102] is behind schedule and over budget. Similarly, some people felt that decommissioning and waste costs are unknown, since they felt the UK has no experience of decommissioning and waste management. Other respondents noted that the record of nuclear power construction cost management in the UK is not good but that any future practice will almost inevitably improve on this record. This is because large project management techniques have improved over the past 25 years and the prospect of genuine international tendering should restrain costs. Against this, respondents noted the novelty of reactor designs that might be used in the UK, with issues around standardised designs and programme build and the political and regulatory risks attached to new designs. These uncertainties could mean that the chances of cost overruns are higher than the chances of achieving cost savings.

2.44    Those who did not agree with the Government's view raised a range of other points:
   - The apparent absence of a mechanism to ensure that nuclear operators maintain insurance or other financial security to cover their liabilities. Some estimated that the absence of such a mechanism reduced the costs of nuclear power by 70%

---

101 For example, see Science Policy Research Unit (SPRU) report.
102 *Nuclear industry revival hits roadblocks*, NewScientist.com, 1 July 2007.

- Notwithstanding the Government's stated intention to ensure that operators built up funds to cover the costs of decommissioning, some respondents were sceptical that operators would actually accumulate sufficient funds
- Costs do not take account of the $CO_2$ produced during construction of the nuclear plant. Construction of a new nuclear power station can require more energy and produce more carbon dioxide than any carbon benefits during operation
- There is no guarantee that a carbon price signal will endure for the life of the nuclear power station or that the carbon trading scheme will not be radically altered in future years
- The costs of offsetting carbon emissions from nuclear power could be higher than from renewables because it would cost more to generate nuclear electricity
- A single accident would render the Government's conservative estimates false and would mean decommissioning of plants before energy companies had acquired the funds to do so.

## Government response

### Cost benefit analysis methodology

2.45 In order to address clearly criticisms of the methodology the Government used in analysing the costs, we have set out below the details underpinning our analysis. We undertook a high level assessment of the costs of nuclear power and other conventional forms of electricity generation. This allowed us to consider the competitiveness (relative to fossil fuels and other low-carbon technologies) of nuclear power generation as an option in the UK's electricity generation mix. We set out in our consultation document that the costs and economics of any new nuclear power station will depend on, among other things, the contracts into which investors enter for the construction of the power station, the cost of capital and ultimately the value attached to the electricity generated.

2.46 The Energy Review Report published in July 2006[103] included a detailed annex on the modelling of the relative costs of electricity generating technologies. The levelised cost of generating electricity[104], which underpins the cost benefit analysis, was calculated by adding the capital costs and the back-end costs onto the operating costs and dividing this by the amount of electricity the plant is expected to generate during its lifetime. The costs are annuitised (made into an annual cost) using a real discount rate of 10% for our "central" case, bounded by 7% in our "low" scenario and 12% in our "high" scenario. The levelised costs are particularly sensitive to the discount rate. By running low and high scenarios, as well as a central case, we assume that differing levels of risk for the private investor and technology maturity are covered. We did not vary the discount rate for different technologies: this would have meant the Government taking a view on private investors' risk profile and their expected return on their investment and would have resulted

103 Energy Review, *The Energy Challenge*, July 2006.
104 The cost per megawatt hour of electricity generated.

in a large number of scenarios for all the different potential profiles. Our discount rate assumptions are not dissimilar to those used in other studies by organisations such as the IEA of the cost of generating electricity[105]. The assumptions on discount rates were based on the IEA study as well as a range of other studies on nuclear costs.

2.47 In order to calculate the welfare benefits attached to reduced emissions of $CO_2$ and enhanced security of supply, we sought to understand the impact on society. The costs of generation from gas and nuclear power were compared and the differences discounted over the lifetime of the plants. In order to calculate the impact on society we used the UK Government's discount rate, as set out in the Treasury Green Book[106]. This is 3.5% for a period up to 30 years in the future and 3% from year 31 to year 75.

## Costs and discount rates

2.48 The Government's cost-benefit analysis used a central assumption on construction cost of £1,250/kW, based on a range of studies, plus costs for interest during construction and onsite waste storage. This gives a total cost of £2.8 bn to build a first of a kind plant with a capacity of 1.6 GW. Sensitivities were also modelled for a lower cost of £850/kW (equivalent to £2.0 bn total build cost) and two higher cost scenarios of £1,400/kW and £1,625/kW (equivalent to build costs of £3.1 bn and £3.6 bn respectively). The total build cost in each scenario is higher than the fixed price contract estimates for the nuclear power station is currently under construction at Olkiluoto in Finland – the cost of which is currently projected to be £2.7 bn.

2.49 While there have been cost overruns and delays in constructing nuclear power stations, such as at Olkiluoto in Finland, experience elsewhere in Europe is different. For example, plants have been built to schedule in France and Romania. Using the central cost assumptions and varying the discount rate results in a levelised cost of £31/MWh (for a 7% discount rate), £38/MWh (for a 10% discount rate) and £42/MWh (for a 12% discount rate). These levelised costs are all higher than the average estimates made by energy companies of £30/MWh[107]. Making different assumptions on factors such as capital costs, carbon prices and discount rates will yield different cost estimates for low-carbon generation technologies but we believe the estimates we have made are prudent and appropriate.

2.50 We have used the higher end of vendors' estimates for the underlying assumptions of the size of the fund needed to pay for back-end costs, including the costs of decommissioning and waste management which we discuss in greater detail at paragraphs 3.46-3.78. Most of these back-end costs will not be incurred until the power station ceases to operate. Under our modelling assumptions therefore generators can

105 For example IEA/NEA, *Projected Costs of Generating Electricity*, 2005 Update; PB *Power, Powering the Nation; A review of the costs of generating electricity*, June 2006.
106 HM Treasury, *The Green Book – Appraisal and Evaluation in Central Government*.
107 The Government's estimates from the cost benefit analysis and the corresponding private sector estimates are set out on pages 66 – 70 and in Figure 4.2 on page 67 of the nuclear consultation document.

accrue the necessary funds over the life of the power station, which we assumed to be 40 years in our analysis. As a part of the total levelised cost of new nuclear power these back-end costs are relatively small, although there is significant uncertainty attached to them. We also make a conservative assumption of 2.2% (in real terms) on the rate at which the fund grows over the accrual period. As part of the work to put in place, through legislation, a robust mechanism to ensure that operators cover their full costs, the Government is undertaking an in-depth exercise to model these back-end costs of waste management and decommissioning. More detail on this work is set out in Section 3. However, in advance of the outputs of this modelling work, we have explored whether higher back-end costs would materially alter the levelised cost and have satisfied ourselves that they do not. Modelling a larger fund size requirement (if the back-end cost is expected to be higher) demonstrates a minimal impact on the levelised cost. In our consultation document, we increased the required fund size by 50% for decommissioning (from £636m to £950m[108]) and retained a 2.2% fund growth assumption, which increased the levelised cost estimate by only £0.3/MWh. Similarly, for waste management we increased the fund size by 15% (from £276m to £320m) which increased the levelised cost by only £0.01/MWh.

2.51 All new generation plants will need connection to the national grid. Our analysis of nuclear generation costs has taken account of grid connection issues[109]. On the level of "spinning reserve" (i.e. generation capacity which can come on stream at short notice) required, some renewable generation technologies, for example wind generation, require significant spinning reserve where intermittency is a key issue. Our analysis takes account of the differences between nuclear and other generation technologies.

2.52 The risk of cost over-runs in construction and the choice of the appropriate cost of capital are genuinely uncertain factors and are best left for the market to determine. Our analysis, based on conservative assumptions, is that nuclear power should be economic in most scenarios we have modelled and particularly where there is a price put on carbon emissions. Where there is a zero carbon price, nuclear provides a positive welfare balance under as many scenarios as it does a negative balance. However, whether energy companies choose to invest in new nuclear power stations is, ultimately, a matter for them.

## Insurance

2.53 Nuclear operators have to maintain insurance or other financial security to cover liability for personal injury and property damage under the Nuclear Installations Act 1965[110]. We intend to consult on amending

---

108 Note these figures are indicative only and should not be taken as a basis for making investment decisions.
109 The issue of grid connection was raised by Greenpeace in their report *The New Rush for Nuclear: An Expensive White Elephant,* November 2007.
110 This implements the special international regime set out in the Paris Convention on Third Party Liability in the Field of Nuclear Energy of 29 July 1960 and the Brussels Supplementary Convention of 31 January 1963, regulating liability for personal injury and third party property damage caused by incidents involving nuclear matter in the course of carriage to or from, or on a licensed site.

this Act to include new heads of liability, such as the cost of measures of reinstatement of impaired environment, and the requirement for insurance or other financial security will also then be extended to cover these new liabilities. As mentioned in the consultation document[111], in accordance with our international commitments, there will continue to be certain potential liabilities that may fall to the Government as a result of a nuclear event.

## Meeting the costs of decommissioning

2.54 We recognise the need for clarity on how operators meet the full costs of decommissioning and their full share of waste management and disposal costs. Section 3 (paragraphs 3.46-3.78) sets out the Government's proposals to ensure that operators of new nuclear power stations meet the costs of cleaning up new power stations when they are decommissioned. It is the Government's policy that the operators of new nuclear power stations must set aside funds over the operating life of the power station to cover the full costs of decommissioning and their full share of waste costs. These financing arrangements must be robust, and designed to deliver sufficient funds even under scenarios such as the insolvency of the energy company or the early closure of the power station. The Government is taking powers through the Energy Bill to introduce this financing mechanism.

2.55 The UK has extensive experience of decommissioning, clean-up and waste management of nuclear facilities. Where necessary, we can also draw upon expertise from overseas. Waste management is a long-term process, being delivered through the MRWS programme. This is described in more detail at paragraph 2.153.

2.56 Our analysis of the economics of nuclear power takes into account all these costs.

## Carbon emissions

2.57 The Government has examined the evidence on carbon dioxide emissions from nuclear plants throughout their lifecycle and has compared them with other forms of generation. As we discussed in paragraphs 2.17-2.20, apart from one study, all the evidence indicates that carbon dioxide emissions from nuclear power stations are very low and at least comparable with emissions from renewable technologies and are considerably lower than generation using fossil fuels.

2.58 Our analysis in the consultation document[112] shows that in the central case and high cases for gas prices and a central case for nuclear costs, nuclear power provides economic benefit regardless of the carbon price.

111 See paragraphs 4.12-4.15 and 4.32 of The Future of Nuclear Power, *The Role of Nuclear Power in a Low Carbon UK Economy*, Consultation Document, URN 07/970, May 2007. Articles 48 and 53 of the Euratom Basic Safety Directive, have now been implemented in the UK.

112 Table 4.4, page 73, The Future of Nuclear Power, *The Role of Nuclear Power in a Low Carbon UK Economy, Consultation Document*, URN 07/970, May 2007.

# Chart 2: Marginal abatement cost curve 2020 with new shadow price of carbon (SPC)

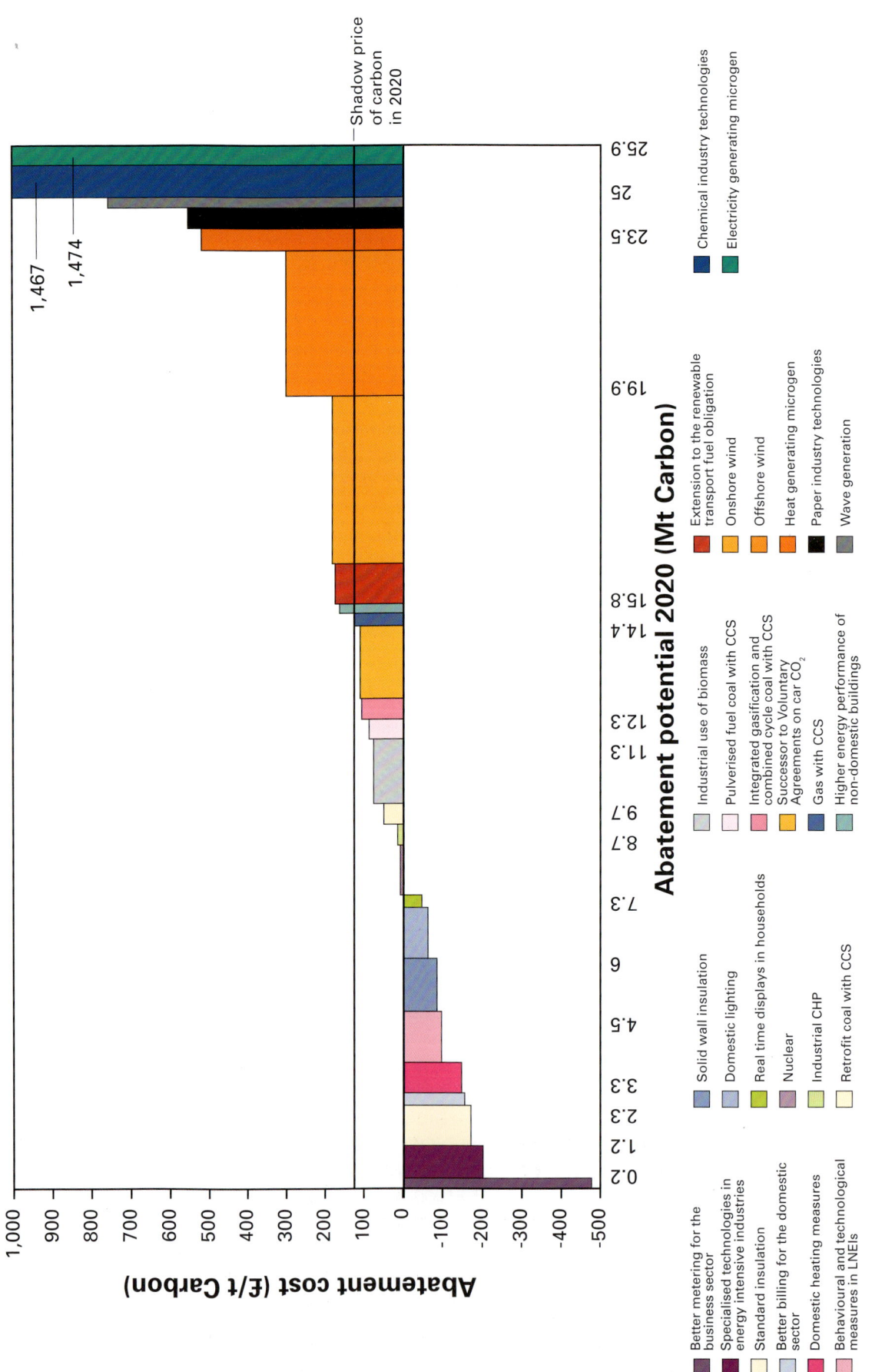

2.59    The cost-benefit analysis of nuclear power[113] provides our analysis
        on the cost effectiveness of alternative low-carbon generation
        technologies. On the basis of the assumptions we used, nuclear power
        is the most cost effective low-carbon generation technology. It has an
        estimated abatement cost of £1/t Carbon (equivalent to £0.3/tCO$_2$) (see
        Chart 2) compared with onshore wind power, the next nearest currently
        available low-carbon electricity generation technology, which has an
        estimated abatement cost of £182/t Carbon (equivalent to £50/tCO$_2$).

## Giving a carbon price signal

2.60    In August 2007, Defra published guidance on the shadow price of
        carbon[114] (SPC) which replaces the social cost of carbon previously
        used in the Government's policy appraisal. The shadow price of carbon
        captures the worldwide costs of the damage due to climate change
        caused by each additional tonne of greenhouse gas emitted. In the
        calculation of the welfare balance, we assumed a carbon price range
        of between €0 and €36/tCO$_2$, the shadow price of carbon is £24/tCO$_2$
        (€34) in 2006 (which is within the range we have assumed for the
        carbon price). Therefore, our view on the welfare balance of nuclear
        generation remains unchanged.

2.61    The UK recognises the importance of a clear framework for carbon
        pricing for new investment in low-carbon technologies, including nuclear
        power. We believe this framework is best achieved at international and
        EU levels.

2.62    The UK, together with our EU partners, will therefore continue to press
        for a broad mandate to the UN's negotiations on a framework for
        climate change. This will include a shared vision to reduce greenhouse
        gas emissions, with the EU being prepared to reduce emissions to 30%
        below 1990 levels by 2020 in the event of an international agreement.
        We will also push for a global carbon market that creates an effective
        global carbon price, and which includes enhanced mechanisms for
        channelling clean development resources to developing countries.
        With growing international consensus on the need to tackle climate
        change, the recent Bali conference agreed a timeframe for international
        negotiations to end by 2009. Amongst other things, these will
        consider opportunities for using carbon markets to enhance the cost-
        effectiveness of tackling climate change.

2.63    The EU ETS is the key plank of the EU's climate change policy. The
        EU ETS sets a cross-EU cap on carbon dioxide emissions, covering
        the electricity generation sector and other major emitters of CO$_2$. The
        scheme is being reviewed in the light of the experience of the 3 year
        phase which ran from 2005 to 2007; and tighter caps have already been
        set for the second phase (2008-2012). A new draft Directive is expected
        from the Commission in January 2008.

2.64    We know that we need further clarity on the number of allowances
        and the level of auctioning to be applied from 2013 and beyond. The

113 Table 9, page 26, *Nuclear Power Generation Cost Benefit Analysis.*
114 DEFRA, *How to use the Shadow Price of Carbon in policy appraisal.*

A White Paper on Nuclear Power

UK wants to see a cap set at EU level, clearly linked to the EU's target to reduce $CO_2$ emissions by 2020, to give a strong, long term signal as to the level of emissions reduction required, with clarity provided on the level of caps in future phases for at least 15 years. We also want to see a significant increase in auctioning for electricity generation. Negotiations on the Directive will take place in 2008 and 2009.

2.65 The UK is committed to working towards strengthening the EU ETS in order to build investor confidence in the existence of a long-term multilateral carbon price signal. We will keep open the option of further measures to reinforce the operation of the EU ETS in the UK should this be necessary to provide greater certainty for investors.

## Risk of accident

2.66 As we set out in the cost-benefit analysis, we have not estimated a monetary value that might be associated with potential accidents. Evidence suggests that the likelihood of such accidents is negligible, particularly in the UK context. Though we cannot dismiss the risk of accidents, we have taken the view that this can be managed through arrangements for design and regulatory and corporate governance for the nuclear industry[115].

## *Our conclusion*

**We have reviewed the arguments and evidence put forward, and based on the conservative analysis of the economics of nuclear power, the Government concludes that, under the most likely scenarios for gas and carbon prices, nuclear power would yield economic benefits to the UK in terms of reduced emissions of $CO_2$ and improved security of supply. It is for investors to determine whether the financing characteristics of nuclear power provide sufficiently attractive returns. However, on the basis of our cost-benefit analysis, we believe that nuclear power is likely to be an attractive economic proposition to them.**

**The Government is committed to working to strengthen the EU's Emissions Trading Scheme (EU ETS) and to building investor confidence in a long-term multilateral carbon price signal. We will keep open the option of introducing further measures to reinforce the operation of the EU ETS in the UK should this be necessary to provide greater certainty for investors.**

---

115 The cost-benefit analysis assessed the probability of a major nuclear accident in the UK and the associated monetary cost.

# The value of having low-carbon electricity generation: nuclear power and the alternatives

**THE GOVERNMENT'S PRELIMINARY VIEW**

The Government believes that given the wide range of uncertainties it is difficult to predict with certainty the future need for and use of energy and electricity.

We have modelled a number of different future scenarios as part of the analysis to support the Energy White Paper. The modelling indicates that it might be possible under certain assumptions, to reduce the UK's carbon emissions by 60% by 2050 without new nuclear power stations. However, if we were to plan on this basis, we would be in danger of not meeting our policy goals:

- Security of supply: we would be reliant on a more limited number of technologies to achieve our goals, some of which (e.g. carbon capture and storage) are yet to be proven on a commercial scale with power generation. This would expose the UK to greater security of supply risks, because our electricity supplies would probably be less diverse as a result of excluding nuclear; and
- reducing carbon emissions: by removing one of the currently more cost-effective low carbon options, we would increase the risk of failing to meet our long term carbon reduction goal.

By excluding nuclear as an option, our modelling also indicates that meeting our carbon emissions reduction goal would be more expensive.

Therefore, the Government believes that giving energy companies the option of investing in new nuclear power stations lowers the costs and risks associated with achieving our energy goals to tackle climate change and ensure energy security.

**Question 5**
**Do you agree or disagree with the Government's views on the value of having nuclear power as an option? What are your reasons? Are there any significant considerations that you believe are missing? If so, what are they?**

## Key arguments and issues presented in responses

2.67    Many people answering this question argued that investment in measures to promote energy efficiency and education would be of far more use than investment in nuclear, and would put the UK in a better position to respond to future uncertainties. Conversely, some respondents suggested that having many diverse power sources would not necessarily increase security of supply. They argued that renewable technologies present lower risk, particularly when it comes to security and safety, are more sustainable and that renewables could deliver

$CO_2$ benefits sooner, and should, therefore, be the focus of future investment.

2.68 Some respondents thought that investing in nuclear would divert investment away from the renewables market. They went on to argue that the Government's cost estimates do not take account of this. They argued that excluding nuclear power would give an impetus to investment in alternatives such as wave and tidal technology, wind, biomass, CCS and energy conservation. Some people suggested that we needed to transform our electricity grid into a smart distributed network which relies on local electricity generation. This would enable the capture of co-generated heat (combined heat and power (CHP)) and also to reduce grid transmission losses. Others made the point that large-scale technologies, such as nuclear, lock the UK into a centralised generation system, thus frustrating the development of local CHP programmes.

2.69 Some people said that the Government would be forced to act as provider of last resort to support investment in nuclear power stations. They argued that nuclear power would have little impact until around 2020 when the Government assumes that it would be the cheapest low-carbon form of electricity generation. They felt this assumption was doubtful in view of the large amount of renewable capacity coming on stream and that, by then, new renewable technologies such as wave, tidal and solar photovoltaic will have matured.

2.70 Some respondents had concerns over the safety and security of nuclear power plants. The threat of terrorist attacks and safety concerns with regards to accidents was highlighted. Some argued that these types of risks made nuclear power fundamentally different from other types of energy. How the Government takes these issues into account is outlined in more detail in relation to Question 6.

## Government response

2.71 The consultation document noted that the UK could meet $CO_2$ emissions reduction targets (60% reduction from 1990 levels by 2050) without nuclear power, but at increased cost and at the risk of missing such targets. In assessing the arguments put forward, therefore, the key point was not whether we can meet targets without nuclear power, but the extent to which it is possible to meet those targets in a practicable manner at low risk and at reasonable cost. Our analysis of how the UK could deliver its energy policy goals without nuclear power is set out in Annex A. We recognise that a future where nuclear power is not an option and energy conservation and alternative technologies are promoted is a scenario supported by some participants in the consultation. Annex A uses analysis originally conducted for the Energy White Paper and the nuclear consultation document to examine the implications for our energy goals if we ruled out the option of new nuclear power stations[116].

116 Energy White Paper, *Meeting the Energy Challenge*, URN 07/1006, May 2007.

2.72   We agree that there are many options that can help us to achieve our energy and climate change goals. However, the future is uncertain, and we cannot say today which options will be the most cost-effective in 50 years' time. Future cost-effectiveness will be driven by the level and structure of energy demand, the rate of technological change and the availability and cost of energy supplies. Because the future is uncertain, the Government's view is that we need a range of different options that are consistent with our targets to reduce carbon emissions reductions and maintain security of supply.

2.73   In particular, we have looked at the potential contribution that we could realise from energy efficiency, and the potential for investment in different low-carbon technologies, such as renewables and CCS. We conclude that investment in nuclear would not prevent investment in other kinds of generating capacity, but that, even if nuclear power were available, we would still need to invest more in energy efficiency and in other technologies. So we are not faced with a choice between nuclear power on the one hand and other technologies and energy efficiency on the other: investment in nuclear would be one of a number of non-mutually exclusive options we would have available to meet our goals.

2.74   Measures to improve energy efficiency already form a key part of the Government's climate change policies. In fact, we believe energy efficiency measures are amongst the most cost effective ways of reducing energy demand and hence carbon dioxide emissions (see the marginal cost abatement curve in Chart 2). Based on existing measures we already expect the energy efficiency of our economy to improve by a third by 2020. Nonetheless, there are limitations to the potential contribution that energy efficiency measures can make in delivering our energy goals. For example households may use the financial rewards from improving energy efficiency to increase their use of energy in other areas, a phenomenon described as the rebound effect; and people might opt to improve their level of comfort, by increasing the temperature at which they heat their home, or by purchasing more energy-consuming products, thereby increasing carbon dioxide emissions[117].

2.75   Further, we will still need a substantial amount of investment in new generation capacity, as we need to replace around a third of our current electricity generation capacity over the next 20 years or so. If we are to meet our long-term $CO_2$ reduction objectives this new generation capacity must be increasingly low-carbon. Energy companies could invest in a number of low-carbon technologies that could help meet our electricity needs whilst reducing the UK's carbon dioxide emissions from the electricity sector. These include renewables, fossil fuels with CCS, and nuclear power. We believe we will need all of these types of generation to play a part in the long-term if we are to achieve our targets while keeping costs down and minimising risks[118].

117 *The Rebound Effect: an assessment of the evidence for economy-wide energy savings from improved energy efficiency*, UKERC, October 2007.
118 These different energy sources are in fact not directly comparable: nuclear is used for base load generation, whereas coal CCS can be used both as baseload and flexible generation, some types of renewables, such as wind, are intermittent, i.e. variable but not predictable, and others, such as tidal, are variable but predictable, and therefore cannot be relied upon at times of peak demand.

2.76 Some observers have suggested that giving energy companies the option to invest in nuclear power would displace investment in renewables. The Government rejects this argument. Energy companies tend to want a diverse portfolio of technologies so as to help them reduce their exposure to changes in the costs of other generation technologies, for example driven by fossil fuel prices. Companies will invest in new power stations on the basis of the expected profitability of those investments. For example, investment in renewables is driven primarily by four key factors: (i) investors' expectations of future electricity prices[119] and of the carbon price; (ii) investors' expectations of the cost of renewables; (iii) investors' expectations of the amount of subsidy available and (iv) any technical barriers to investment, such as planning or access to the grid. We believe that allowing energy companies to build new nuclear power stations will have at most marginal impact on any of these factors. Thus, if both renewables and nuclear were available as options, energy companies may want to have both.

2.77 Investment in renewable electricity is also strongly driven by the Renewables Obligation (RO)[120]. This requires electricity suppliers to obtain an increasing proportion of their electricity from renewable sources, from around 5% today to 15.4% by 2015 or pay a buy-out price for any shortfall[121]. Since its inception in 2002 the RO has promoted an increase in renewables generation from 1.32% of the mix to 4.43% in 2006, and there are currently 1.3 GW of renewables capacity under construction and around 9.5 GW consented. The Government has consulted on how to make the obligation more cost effective by providing more support to technologies that are further from the market and less to technologies that are close to being competitive with generation from fossil fuels. The Government intends, subject to Parliamentary approval and State Aid clearance from the European Commission, to "band" the Renewables Obligation in this way from 1 April 2009[122].

2.78 The Government also helped to secure agreement with other EU member states at the 2007 Spring European Council to a binding target of 20% of the EU's energy consumption to come from renewable sources by 2020. This will cover electricity, heat and transport fuels. The European Commission is expected to publish early this year a draft directive to implement this target, including the contributions to be made from each Member State. This will then be subject to negotiation, with a final decision expected in early 2009. Therefore we do not yet know what the UK contribution will be, but it is clear that we will need to raise significantly the proportion of our energy, including electricity, that comes from renewable sources. We will launch soon a consultation on how we are to achieve our targets and we will publish our full renewable energy strategy once the EU Directive has been agreed.

119 In the UK given our policy regime; in countries with feed-in tariffs, for example, investors are immune to future electricity prices.
120 http://www.berr.gov.uk/energy/sources/renewables/policy/renewables-obligation/page15630.html
121 Our current estimates suggest that the Renewables Obligation will cost over £1.5 bn per annum by 2015.
122 Further details can be found in *Renewable Energy: Reform of the Renewables Obligation*, URN 07/636, May 2007, published with the Energy White Paper.

2.79 The Government accepts that, given the long timescales for planning and construction, new nuclear power generation may not make a large contribution to our targets to reduce $CO_2$ emissions by 2020. However, our targets are not limited to 2020: a further goal is to reduce $CO_2$ emissions by 60% by 2050. Indeed, if we are to meet our long-term targets for $CO_2$, this will mean that both nuclear and renewables technologies could have a significant share of the market, together with fossil fuel generation coupled with CCS (assuming that CCS[123] proves to be a viable technology). On this basis, the Government does not believe that investing in nuclear power will damage the prospects for renewable electricity.

2.80 By 2050 it is possible that most new fossil-fuelled power stations will be able to deploy CCS technology. That could reduce carbon emissions from burning fossil fuels by up to 90% compared to today. The Government supports the development of CCS, and recently launched a competition for the development of a demonstration project[124]. Successful development would provide the UK with cleaner fossil-fuelled energy, so that in the future CCS and nuclear can complement each other as low-carbon sources of electricity. However, CCS is an as yet unproven technology and we have to acknowledge that there is some risk that safe and reliable CCS for power generation might not be proven or deployable at scale and at reasonable costs. This could happen if the projected costs turn out to be too high or if it proves difficult to develop safe ways to transport and store $CO_2$.

2.81 We also recognise that there is value in having decentralised electricity generation. Connecting electricity generation closer to the point of use reduces the extent of the infrastructure needed to transport electricity. This suggests that there could be costs savings and lower losses of electricity during its transportation to the customer. However, whilst our findings suggest[125] that some distributed generation may be economically competitive with centralised generation, the overall costs of generating our future electricity are likely to be lower if we retain a framework where Distributed Generation is a complement rather than an alternative to centralised generation, be it nuclear, renewables, coal or gas-fired power stations. In this context, we believe that new nuclear power stations would not necessarily affect the potential from Distributed Generation.

2.82 Even though increased energy efficiency, use of renewables, CCS and Distributed Generation will be vital if the UK is to achieve its emissions targets, we have to consider the limitations and risks that are associated with these options. Those risks and limitations will still be there if nuclear power is a part of the UK's energy mix, but as we show in Annex A, they would be more pronounced and magnified if the option of building new nuclear power stations were ruled out. That would mean that it is likely to be more expensive to achieve our goals if nuclear power is not an option.

123 CCS involves capturing the carbon dioxide emitted when burning fossil fuels (as much as 90% of the volume), transporting it and storing it in secure spaces such as geological formations, including old oil and gas fields and aquifers under the sea bed.

124 See http://www.berr.gov.uk/energy/sources/sustainable/carbon-abatement-tech/ccs-demo/page40961.html.

125 See http://www.berr.gov.uk/energy/whitepaper/consultations/distributed-generation/page39557.html.

## *Our conclusion*

Having reviewed the arguments and evidence put forward, the Government believes that giving energy companies the option to invest in new nuclear power stations reduces the costs and risks associated with tackling climate change and ensuring energy security. Nuclear power needs to be part of an overall approach to electricity generation. We will also take further steps to support renewables, Carbon Capture and Storage and Distributed Generation as outlined in the Energy White Paper and implemented through the Energy Bill.

# Safety and security of nuclear power

**THE GOVERNMENT'S PRELIMINARY VIEW**

Based on the advice of the independent nuclear regulators, and the advances in the designs of nuclear power stations that might be proposed by energy companies, the Government believes that the safety, security, health and non-proliferation risks of new nuclear power stations are very small and that there is an effective regulatory framework in place that ensures that these risks are minimised and sensibly managed by industry. Therefore, the Government believes that they do not provide a reason to prevent energy companies from investing in new nuclear power stations.

**Question 6**
**Do you agree or disagree with the Government's views on the safety, security, health or non-proliferation issues? What are your reasons? Are there any significant considerations that you believe are missing? If so, what are they?**

## *Key arguments and issues presented in responses*

2.83  Whilst some respondents agreed with the Government's views on the safety and security of nuclear power, others referred to incidents or new studies[126] which, they believed raised issues in relation to safety, security or health. In general, many respondents voiced a high level of concern about safety issues.

2.84  A number of respondents questioned the ability of the Health and Safety Executive (HSE), the safety regulator, to cope with an increased workload if there were to be new nuclear power stations. Others questioned whether the private sector would sacrifice health and safety standards in the pursuit of profits.

2.85  Respondents raised a number of broad issues on security. These ranged from the security of any new nuclear plant to the security of the transportation of materials associated with the operation of nuclear plants. Little of this appeared to relate directly to new nuclear power stations but instead was raised as a concern associated with nuclear power stations more generally. Respondents questioned whether the Office for Civil Nuclear Security (OCNS), now part of the HSE, could be fully satisfied that all vehicles entering or leaving sites could be sufficiently checked for unauthorised material or personnel. Some were concerned that established 'no-fly zones' around nuclear sites would not be effective in preventing aircraft from crashing into nuclear sites[127]. There were also concerns about the ability and the accuracy of the intelligence services to evaluate and identify the level of any

---

126 See paragraphs 2.106 and 2.107.
127 No fly zones are a safety feature to allow sufficient time for a small plane to glide clear. A jet plane crosses a no fly zone in about six seconds. They are therefore not designed as a security measure but as a safety one.

terrorist threat to the civil nuclear industry, and about possible terrorist infiltration. Others were worried about how increased security demands would affect safety and security regulators and asked whether any new nuclear programme would lead to an increase in the budget and size of the Civil Nuclear Constabulary (CNC).

2.86 On health issues, respondents questioned the long-term effect of living near nuclear power stations. Similar concerns were raised about the accuracy of studies on cancer clusters[128]. A number of respondents raised more general comments on the Government's policy on safe levels of radiation exposure and challenged the Government's position on basing safe levels of exposure on ICRP[129] research data.

## Government response

### Safety and security

2.87 The UK has strict, independent, safety and environment protection regimes for nuclear power which fulfil the requirements of the Euratom Treaty with regard to radiation protection[130]. Any new nuclear power station will be subject to safety licensing conditions and will have to comply with the safety and environmental conditions set by the regulators in their licences and authorisations. As we said in our consultation document, new nuclear power stations will need authorisation from the relevant environmental agency under the Radioactive Substances Act 1993 before making any discharges of radioactivity. Statutory obligations require that radiation exposures not only comply with dose limits but are as low as reasonably achievable (ALARA). The environment agencies will ensure that radiation exposure of members of the public from disposals of radioactive waste, including discharges, are ALARA by requiring new nuclear installations to use the best available techniques (BAT) to meet high environmental standards. This will help ensure that radioactive wastes created and discharges from any new UK nuclear power stations are minimised and do not exceed those of comparable power stations across the world.

2.88 Some respondents referred to incidents at nuclear power stations in Sweden and Germany as examples of the potential safety issues associated with nuclear power plants. In the cited Swedish example the Swedish regulator found that the incident at Forsmark[131] did not result in any consequences to the public or the environment and that no radiological release occurred. They noted that the handling of the event went according to simulated routines, working practice and emergency procedures. Likewise two emergency shut downs at plants in Germany

---

128 Professor B. A. Bridges, Committee on Medical Aspects of Radiation in the Environment (COMARE), 10th Report, *The incidence of childhood cancer around nuclear installations in Great Britain* and Professor A. Elliott, Committee on Medical Aspects of Radiation in the Environment (COMARE), 11th Report, *The distribution of childhood leukaemia and other childhood cancers in Great Britain 1969 – 1993*.
129 International Commission on Radiological Protection.
130 Council Directive 96/29/Euratom of 13 May 1996, laying down the basic safety standards for the health protection of the workforce and general public against the dangers of ionising radiation, Official Journal of the European Communities (L159 29.6.1966, p.1).
131 http://www.ski.se/dynamaster/file_archive/060914/dca864e98cb5363d39cc03ac0b29f1ee/SKI%20ver%20 %20Executive%20summary%20SKI%20review%20of%20F1%20startup%20reques..pdf

Department for Business, Enterprise and Regulatory Reform MEETING THE ENERGY CHALLENGE

(Brunsbüttel and Kruemmel) went according to plan with the shut downs being executed without risk to the environment or personnel. In the UK there have been a number of incidents at nuclear sites (including those undergoing decommissioning programmes) where there have been breaches of regulations that have resulted in prosecutions[132]. Whilst these breaches have occurred we are confident that UK's regulatory structure will ensure that should similar events occur, the systems and processes will be in place to minimise the risk of harm to people or the environment.

2.89 A number of respondents referred to incidents at Three Mile Island in Pennsylvania in 1979 and the Chernobyl nuclear power station in the Ukraine in 1986. However, for a number of reasons we must be careful before comparing past accidents that happened abroad with anything that might occur at new civil nuclear power stations in the UK. In particular, regulatory scrutiny of reactor operations in the former USSR was far less rigorous than it is in the UK today. We must also remember that many of these past accidents occurred in power stations with designs that would not be acceptable to regulators in the UK.

2.90 We noted that concerns were expressed during the consultation that the private sector would put profits before safety. Safety, security and environmental protection must be a priority for any operator irrespective of whether operators of new nuclear power stations are in the private or public sector. The regulatory system does not, nor should it, distinguish between operators in the private and public sectors. Maintaining high safety and environment protection standards, minimising operational upsets and avoiding unplanned operational shut downs are essential to continuous operation and therefore the profitable performance of the power station. We do not therefore see a conflict between safety and profitability.

2.91 The UK regulatory regime is based upon the principle of independent regulators backed up by tough sanctions. The Government believes that the regulatory process is capable of overseeing existing facilities as well as any new nuclear power stations irrespective of who owns and operates them. A recent review by the IAEA concluded that the HSE's regulatory arrangements are mature and transparent, with highly trained and experienced inspectors[133]. Whilst there can be no room for complacency, the UK has a strong safety record with no events relating to a civil nuclear power station with off-site consequences or where all the safety barriers that are an inherent part of the design were breached.

2.92 Some concerns were raised in the consultation about whether the HSE's Nuclear Installations Inspectorate (NII) would have sufficient resource to deal with new nuclear licensing. Others questioned the inspectorate's effectiveness in scrutinising the reactor designs that are likely to form part of a worldwide fleet rather than, as in the past, UK-specific designs and emphasised the importance of standardised

132 Such as the prosecution of the UKAEA for breaches of the Health and Safety Act in July 2007 at Dounreay.
133 *Integrated Regulatory Review Service (IRRS)*, IAEA, April 2006.

designs. We welcome the preparations and approach that the nuclear regulators have made to secure nuclear safety, security and environmental protection in the advent of new nuclear build in the UK. This includes development of the GDA process, enhanced governance and project oversight arrangements and joint working between the regulators.

2.93   To ensure that the NII can deal effectively with the challenges of a new build programme, the Government has recently worked to ensure that the NII has the ability to recruit and retain the calibre and numbers of staff that it needs by authorising the HSE to increase the salary levels of NII's nuclear inspectors. We will work with all the independent regulators to explore ways of enhancing further the transparency and efficiency of the regulatory regime, without diminishing its effectiveness in dealing with the challenges of new build. Details of the regulatory and advisory structure for nuclear power are set out at Annex C.

2.94   The security of civil nuclear material and sites in the UK is regulated by the HSE's OCNS in accordance with relevant national legislation which fully reflects international obligations and guidelines. OCNS regularly inspect site security arrangements and requires operators to improve their systems where it is found to be necessary.

2.95   The Government appreciates the concerns raised regarding the details of the security measures taken to reduce the risk of incidents against nuclear power stations. There are understandable restrictions placed on the publication of the details taken to protect nuclear sites. However, the OCNS places strict obligations on site operators and requires site security plans to be approved by it and for them to be regularly reviewed. To ensure that the appropriate level of security is in place, the OCNS obtains all relevant threat information from the Joint Terrorist Analysis Centre and requires all operators to respond appropriately. Since the "9/11" attacks, the Government has reviewed aviation security, but the extension of no-fly zones around nuclear sites was for safety, not security reasons.

2.96   A minority of respondents raised concerns about the possibility of an aircraft crashing in to a nuclear reactor and the effect of its impact. The Government's position on this remains unchanged from its stated position in the consultation[134] document, namely that modern nuclear installations are designed to withstand the impact of an aircraft and that the OCNS has worked with operators and the NII to develop measures to mitigate the risk of a deliberate large aircraft crash. Additionally the regulatory assessment process requires power station designs to take into account all reasonably foreseeable threats from both natural and man made hazards. This includes meteorological phenomena, the effects of climate and landscape change impacts, geological disturbance, seismic activity, flooding and aircraft impact. The regulators will require evidence that designers have taken proper account of all loadings that the plant may need to withstand as well as the robustness of the safety related structures and equipment.

134 The Future of Nuclear Power, *The Role of Nuclear Power in a Low Carbon UK Economy, Consultation Document*, URN 07/970, May 2007, p 110, paragraphs 6.44 – 6.47.

These will be judged against relevant safety principles and recognised national and international standards.

2.97 The OCNS will ensure that security measures are included in plans for the construction of any new nuclear power station from the outset. Doing so will avoid the need to retrofit security measures once construction is under way. This will also enable regulators to make an early judgement on the most appropriate measures for any construction site[135]. This will help ensure that security is ingrained into practices at any site from day one.

2.98 The threat of infiltration is taken very seriously. For that reason, site operators are required to ensure that anyone accessing nuclear materials is properly vetted. OCNS provides a security vetting service for all permanent employees and all contractors working in the civil nuclear industry. The service complies with, and is governed by, the same nationally agreed standards and procedures which apply to pre-appointment checks and National Security Vetting. Clearances are granted only after the applicant's request has been investigated and has satisfied the criteria appropriate to the level of access required. Clearances are revalidated at agreed intervals, again in line with nationally agreed practice.

2.99 Following a review[136], the Nuclear Industries Security Regulations 2003 were found fit for their current purposes. While the OCNS will continue to keep all parts of the regulatory framework under review to ensure it remains that way, the Government remains confident that the current security arrangements otherwise remain robust and can adapt to the industry's changing circumstances and changes in the threat.

2.100 The Energy Act 2004 created a standalone police force. On 1 April 2005, the CNC took over from the previous UKAEA constabulary the primary role of protecting the UK's civil nuclear sites and nuclear material in transit. The CNC is accountable to an independent Civil Nuclear Police Authority and through that to the Secretary of State for Business, Enterprise and Regulatory Reform. The budget of the CNC is set by the Police Authority: all costs are met by the operators and not by the taxpayer.

2.101 In April 2007, following a review of the role of the OCNS and the operational aspects of the UK Safeguards Office (UKSO), both functions and staff were transferred from the then DTI to HSE as part of the Nuclear Directorate. This means that a single organisation has responsibility for work on safety and security and the oversight of international safeguards.

---

135 *The State of Security in the Civil Nuclear Industry and the Effectiveness of Security Regulation, April 2006 to March 2007*, OCNS.
136 Review of the Nuclear Industries Security Regulations (NISR) 2003, led by the Department of Trade of Industry and which closed on 1 December 2006.

## Non-proliferation

2.102 Some respondents expressed concern about the possibility of diversion of nuclear material and the proliferation of nuclear weapons. All civil nuclear material in the UK is subject to Euratom Safeguards, which are designed to detect the diversion of nuclear material to weapons or any other undeclared use. Existing nuclear operators are required to provide the European Commission with design information on installations and accountancy reports for nuclear materials. The Euratom Treaty[137] also requires that the Commission's inspectors have access at all times to all places, data and personnel in order to verify the safeguards information submitted in order to provide assurance about the non-diversion of nuclear material. Euratom Safeguards will apply to any new nuclear power station in the UK, and the stations will also be liable to IAEA inspections under the terms of the UK safeguards agreement with the IAEA and Euratom.

2.103 Concerns were also raised that new nuclear power stations in the UK would make it harder for the UK to press for the abandonment of nuclear power worldwide in the interests of non-proliferation. The Government does not accept that pressing other countries to forego nuclear power is an effective approach to non-proliferation. Rather, multi-lateral action is needed to support and strengthen the non-proliferation regime through the Nuclear Non Proliferation Treaty, under which signatories which are non-weapons states have a right to the peaceful development of nuclear power.

2.104 A number of respondents raised concerns over reports in the press on unaccounted amounts of radioactive material. There should be no nuclear material unaccounted for (MUF)[138] at any power station, since accountancy of the nuclear materials consists of verifying items of fresh and spent fuel at the site. MUF occurs when nuclear material is processed during fuel fabrication or reprocessing, for example. These plants are subject to the international safeguards regime.

2.105 With regard to new nuclear build, the Sustainable Development Commission[139] has said that new reactors are likely to be unattractive as a source of nuclear proliferation. This is because the design of any new nuclear power stations would require fuel that needs considerable further treatment before it could be used in weapons. We agree with this analysis.

## Health

2.106 In its 11th report, the Committee on Medical Aspects of Radiation in the Environment (COMARE) reported on geographical variations in the incidence of different types of childhood cancers to relate findings around nuclear power stations to the general geographical

137 http://www.iaea.org/Publications/Documents/Infcircs/Others/infcirc263.pdf
138 Further information on nuclear material unaccounted for can be found at http://www.hse.gov.uk/nuclear/safeguards/materials.htm.
139 Sustainable Development Commission, Nuclear Paper 6: *Safety and Security*, March 2006.

epidemiology of childhood cancers[140]. The report, following up its 10th report (COMARE 2005), found that there was no general pattern of increase of these childhood diseases around nuclear power stations. This study, sponsored by the Department of Health, analysed over 32,000 cases of childhood cancers and is believed to be the largest study of its type in the world. Among its recommendations, the report said that the incidence of childhood leukaemia and other cancers in the vicinity of Sellafield and Dounreay was raised, and should be kept under surveillance and under periodic review. The report also recommended that its findings be confirmed by independent research. The Department of Health has accepted these recommendations.

2.107 During the course of our consultation in July 2007, a separate report identified that leukaemia rates were higher in children and young people living near nuclear facilities[141]. However, it concluded that there was no clear explanation for this and that further research is needed before firm conclusions can be drawn from the report. A report was also published by the German Federal Office for Radiation Protection on a study into childhood cancers in the vicinity of nuclear power stations in Germany[142]. The report concluded that whilst in Germany it believes that there is a correlation between the distance of the child's home from the nearest nuclear power station and the risk of developing leukaemia, it did not follow that ionising radiation emitted by German nuclear power stations was the cause. Childhood cancer is also related to socio-economic factors and this does not seem to have been taken into account in the German study. The study also covers a relatively small sample in comparison to COMARE's 11th report which contains 32,000 cases.

2.108 The ICRP has reviewed its position on radiation risks and the Health Protection Agency (HPA) will comment on these recommendations in due course. It is likely that these will include only minor changes to risk estimates and will recommend the continued use of a dose limit of 20mSv/y for workers and 1mSv/y for members of the public. The ICRP's recommendations form the basis of the requirements of both the UK and EU on radiation safety, leading to a high level of international harmonisation. Some groups have challenged whether the internationally recognised 'dose limit' is the correct way to assess the impact of radiation on a person. The Government, in line with other countries believes that using dose limits is the correct way to assess the impact of radiation on individuals.

2.109 As we explained in our consultation document, everyone is exposed to natural background radiation[143]. Most of our exposure, around 80%, comes from natural sources, such as radon gas that emanates from the ground, cosmic rays from outer space and radiation from rocks such as granite. Medical procedures, such as X-rays, account for around

140 Professor A. Elliott, Committee on Medical Aspects of Radiation in the Environment (COMARE), 11th Report, *The distribution of childhood leukaemia and other childhood cancers in Great Britain 1969-1993.*
141 P.J. Baker, D.G. Hoel (2007) *Meta-analysis of standardized incidence and mortality rates of childhood leukemia in proximity to nuclear facilities*, European Journal of Cancer Care 16 (4), pp 355–363.
142 http://www.bfs.de/en/bfs
143 The Future of Nuclear Power, *The Role of Nuclear Power in a Low Carbon UK Economy, Consultation Document*, URN 07/970, May 2007, pp 111-114.

14% of our total annual exposure to man-made radiation. The collective dose to workers from emissions from nuclear power stations is below those experienced by other workers prone to radiation exposure in their workplaces (see Box 2)[144].

## Box 2

| Average annual exposure rates form artificial sources | |
| --- | --- |
| Source | Average annual dose |
| Workers in nuclear industry | 0.4mSv |
| Medicine/research industry | 0.1mSv |
| Aircrew | 2mSv |
| Discharge of liquid radioactive wastes into the marine environment | 0.7µSv (0.0007mSv) (10% from nuclear industry remainder from other sources e.g. phosphate, oil and gas industries) (0.0007mSv) |
| Medical irradiation | 410µSv (0.41mSv) |
| Exposure to airborne discharges from nuclear power stations | 0.1µSv |
| **Average annual exposure rates from naturally occurring sources** | |
| Source | Average dose level |
| Cosmic radiation at ground level | 300µSv (0.3mSv) |
| Terrestrial gamma radiation | 350µSv (0.35mSv) |
| Radon | 1300µSv (1.3mSv) |

## *Our conclusion*

**Having reviewed the arguments and evidence put forward, and based on the advice of the independent regulators, and the advances in the designs of power stations that might be proposed by energy companies, the Government continues to believe that new nuclear power stations would pose very small risks to safety, security, health and proliferation. We also believe that the UK has an effective regulatory framework that ensures that these risks are minimised and sensibly managed by industry.**

144 The Future of Nuclear Power, *The Role of Nuclear Power in a Low Carbon UK Economy, Consultation Document*, URN 07/970, May 2007, Table 6.2, p 114.

# Transport of nuclear materials

**THE GOVERNMENT'S PRELIMINARY VIEW**

Given the safety record for the transport of nuclear materials, the assumption that spent fuel will not be reprocessed and the strict safety and security regulatory framework in place, the Government believes that the risks of transporting nuclear materials are very small and that there is an effective regulatory framework in place that ensures that these risks are minimised and sensibly managed by industry. Therefore, the Government believes that they do not provide a reason to not allow energy companies to invest in new nuclear power stations.

**Question 7**

**Do you agree or disagree with the Government's views on the transport of nuclear materials? What are your reasons? Are there any significant considerations that you believe are missing? If so, what are they?**

## Key arguments and issues presented in responses

2.110 Whilst a number of respondents agreed with the Government's views on the transport of nuclear materials, some were concerned about the associated security arrangements. In particular, they queried the ability of containers used in transit to withstand accidents and suggested that terrorists could intercept material. Some suggested that, in the absence of any meaningful research to challenge engineering and scientific conclusions reached over 30 years ago, there should be a root and branch reassessment of the containers and methods used to transport of nuclear material. A number of respondents cited the incident that occurred in 2002 in respect of a radioactive source from a hospital, incorrectly sealed when being taken to Sellafield[145].

## Government response

2.111 Whilst the first IAEA regulations on the Transport of Radioactive materials date from 1960s, they have been reviewed and updated many times to reflect the latest developments in technological or scientific knowledge[146].

2.112 From 1996 until 2004, the European Commission funded 35 studies on the safety of the transport of radioactive material[147]. They include evaluations of safety of practices in various conditions, technological evaluation of components, development of tools and criteria for evaluation of safety and developments of requirements for specific types of materials. In addition, several international conferences to

---

145 This incident in 2002 involved a radioactive medical device being transported from a hospital.
146 IAEA Safety Standards for protecting people and the environment, *Regulations for the Safe Transport of Radioactive Material*, 2005 edition.
147 http://ec.europa.eu/energy/nuclear/transport/projects_en.htm

review the effectiveness of the transport regulations and operations are held at regular intervals.

2.113 As a result, the Government continues to believe that the relevant regulations and associated enforcement arrangements are robust and provide a high level of safety as demonstrated by the very low impact on the health of the public and workers when radioactive material is transported in compliance with the regulations.

2.114 The Health Protection Agency has conducted an assessment of all events[148] involving radioactive material during transport since 1958 and found that most of the recorded events during this period had not resulted in any significant health effects for workers or members of the public. All 19 significant dose events involved industrial radiography sources that were transported without the source being properly returned to their container and occurred mainly in the 1970s, only two have occurred since the mid -1980s[149].

2.115 The Government believes that, given the strict safety and security regulatory framework in place, the risks of transporting nuclear materials are very small and that there is an effective regulatory framework in place that ensures that these risks are minimised and sensibly managed by industry.

## Our conclusion

**Having reviewed the arguments and evidence put forward, and given the safety record for the transport of nuclear materials and the strict safety and security regulatory framework in place, the Government believes that the risks of transporting nuclear materials are very small and there is an effective regulatory framework in place that ensures that these risks are minimised and sensibly managed by industry. The Government believes that this is not a reason not to allow energy companies to invest in new nuclear power stations.**

148 An event is an occurrence involving administrative errors, improper preparation of packages, physical occurrences such as traffic accidents or damage to the package. The event referred to involved improper preparation of the package and was unrelated to the generation of nuclear power.
149 J.S. Hughes, D. Roberts and S.J. Watson *Review of Events Involving the Transport of Radioactive Materials in the UK, from 1958 to 2004, and their Radiological Consequences*, July 2006.

Department for Business, Enterprise and Regulatory Reform MEETING THE ENERGY CHALLENGE

# Waste and decommissioning

## Key arguments and issues presented in responses

2.116 Many respondents to the consultation, including supporters of new nuclear power stations as well as by those opposed to the idea, raised issues concerning waste management and decommissioning. Significant concerns about waste also featured prominently at the deliberative events we held with the public.

## Opinions on the Government's preliminary view on waste and decommissioning

2.117 There was significant support among respondents for the Government's preliminary view on waste and decommissioning. Several respondents commented that a deep geological disposal facility would be the best technical and long-term solution for managing waste from new nuclear power stations. Some made the point that, if we are to have new nuclear power stations in a responsible way, we need to make progress with implementing the recommendations of the Committee on Radioactive Waste Management (CoRWM) to build a geological disposal facility, commonly known as a repository, to deal with higher activity level waste in the long-term.

2.118 Several respondents said that technology for dealing with radioactive waste already exists and that other countries, such as Finland and the United States of America, are pursuing it. They suggested that the Government should learn from other countries with nuclear power, where arrangements are in place for the longer-term storage or disposal of waste, to make the most of available best practice. Some commented that any outstanding issues concerning the management of waste should not prevent new nuclear build.

2.119 Some respondents felt that the time during which waste would be held in interim storage might allow the development of new technologies, but that even without these, the development of a geological disposal facility would provide sufficient safety. There was some agreement that the best solution for spent fuel is to store it on the site of the power station until a decision is made on the location of the geological disposal facility. Because it usually implied a facility on the surface, some considered that interim storage posed a greater risk than long-term disposal. Some suggested that any geological disposal facility should be able to expand in the future should future generations wish to continue to take advantage of nuclear power.

2.120 Some respondents commented that the UK has built up significant expertise and that the Government should use this expertise to educate the public better. Others felt that in the design and operation of new stations, we could learn from the management of wastes from earlier power stations. Several respondents made the point that modern designs of nuclear power station produce less waste than earlier designs. They also felt that decommissioning should cost less in the future thanks to automation in decommissioning processes and improved reactor design.

2.121 Several respondents identified the UK's good safety record in handling radioactive waste. Some felt that nuclear facilities, including storage sites, are better engineered and far safer than those used to store other industrial waste and heavy metals, so it is disproportionate for the public to be more concerned about safety on nuclear sites. Some participants felt that the regulations dealing with radioactive waste are not consistent, for example, as materials that come into contact with radioactive material in hospitals are not dealt with as securely as waste

from nuclear sites. They felt that other facilities dealing with radioactive materials ought to be brought up to the standards expected of nuclear power stations.

2.122 There was also widespread concern about waste. Many respondents felt that the Government currently lacks a solution for existing nuclear waste and should not compound the problem by creating new waste. Several respondents said that the Government had failed to find a site to dispose of existing waste and doubted that it would find a site for new waste. Some expressed concern that a significant number of unknowns are associated with nuclear waste and argued that we should not expose ourselves to additional unnecessary risk. Instead, they considered it would be better to find ways to meet our targets for $CO_2$ emissions that do not produce further nuclear waste that takes thousands of years to decay. Some argued that burying waste in a geological disposal facility can never be safe, on the basis that the technology is unproven and seismic activities could not be predicted. They doubted the safety of the technology needed to build a repository and questioned what would happen if radioactivity were to leak from a repository, arguing that such a leak would be inevitable at some point in the future. They also questioned the effectiveness of CoRWM's technical studies.

2.123 Some felt that CoRWM did not bring forward a safe way of dealing with nuclear waste, but rather a means that was the best under the circumstances. For this reason, they said, we should continue to research the options. Others suggested that we should store all waste above ground where we can monitor it and so have easy access to tackle unforeseen problems. Some felt concerned that the waste would be buried in a geological disposal facility and forgotten: they wanted reassurance that the waste would continue to be monitored to ensure long-term safety, regardless of political and other changes. A small number of respondents referred to the need to be open to other, possibly better, technical solutions which may come along in the future. For example, they mentioned partitioning and transmutation[150] and fast breeder[151] reactors, which could reduce the volume of the waste and give us access to a greater proportion of the fuel's energy content.

2.124 For some it was important that the management of waste should be led by scientific merit rather than political imperative. Others, however, felt that waste disposal is not a technical problem but a matter of political will. A small number felt it should be left to experts to make decisions on how to manage waste from new nuclear power stations, although they wanted the public to be kept informed of progress, including information on the costs of a geological disposal facility. Others

150 Partitioning and transmutation is a proposed method to separate out individual radionuclides in long-lived radioactive wastes (with half-lives of thousands of years) and to convert them into short-life wastes (with half-lives of tens or hundreds of years). The technique is primarily an area for research and has not been proven to be viable on an industrial scale (see CoRWM Report, *Partitioning and Transmutation*, August 2004).

151 Fast breeder reactors could produce smaller amounts of long-lived high-level radioactive wastes. Research into next or "Fourth Generation" reactors is not expected to produce a commercially viable reactor design until about 2030. Therefore these reactors are not expected to be commercially available before the closure of the UK's operating reactors.

favoured citizens' panels to examine the issues and make decisions on the storage and disposal of new waste.

## Impact of new waste on the management of existing waste

2.125 Several respondents expressed the view that it would be desirable to dispose of both new and legacy waste in the same geological disposal facility. They felt that this would ensure that all such waste is stored where security can be maximised and properly monitored. Some felt that by producing new waste, a programme of new nuclear build would act as a catalyst for developing waste disposal facilities which would make it easier, and cheaper, to deal with existing waste. A small number of people commented that the legacy waste inventory can be calculated with a high level of accuracy so spare capacity could be made available in the geological disposal facility for new waste.

2.126 Many thought that new waste would make little difference as we have to deal with existing waste regardless of whether new nuclear power stations are built. They also observed that with improved design and fuel utilisation, new plants would add relatively little to existing waste volumes.

2.127 Some commented that only by drawing a line under the creation of waste could the Government effectively and efficiently deal with legacy waste. They felt that adding new waste to the inventory could affect the technical parameters of the facility, and not just its size. Several respondents commented that if spent fuel from new nuclear power stations were not reprocessed, this would further increase the volume of higher activity waste that we would have to manage.

## Ethical considerations

2.128 There was some agreement among respondents that there were important ethical issues arising from a decision on whether to create new waste. There was a feeling that the ethical debate was not about the pros and cons of nuclear power, but more about managing the demand and supply of energy in general. This includes the environmental impact of renewables. Some respondents felt that policy should be guided by the scientific and financial feasibility of the proposed solutions and that the ethical issues should not be given as much weight. Other respondents felt that dealing with climate change far outweighs the question of whether or not to proceed with new nuclear power stations. In their view, ethical considerations are heavily weighted towards reducing $CO_2$ emissions and using nuclear as a proven way of doing this. Many respondents felt that, if properly treated, nuclear waste posed a far smaller risk to future generations than the impact of $CO_2$ emissions on climate change. A small number commented that the effect of waste on future generations is not unique to nuclear and that all forms of power generation bear some economic and environmental risks. An example was given of slag from coal-fired power stations.

2.129 Others said we have a responsibility to future generations to use the technology available to us and to build new nuclear power stations. Some respondents raised the ethical issues arising from the need to ensure security of energy supply. These people felt that failing to secure UK energy supplies could have a greater impact on future generations than the impact of climate change. They noted that where energy supplies are cut off, civil breakdown quickly follows.

2.130 Other respondents felt that the balance of argument on ethics is such that new nuclear power stations should not be built. Nuclear waste would be an unacceptable legacy to leave to future generations who may not have the required skills, abilities and resources to look after it. They also observed that we cannot guarantee societal stability for the lifetime of nuclear waste or even for the period in which new nuclear power stations would need to be decommissioned. Others thought that new nuclear waste cannot be justified when there are other ways to tackle climate change such as introducing a carbon tax, increasing energy efficiency and increasing investment in renewables. Some respondents felt that the issue of climate change ought to be kept entirely separate from that of nuclear waste. Their view was that nuclear generation would have a small impact on our $CO_2$ emissions, and that possible benefits do not offset the risks involved in building new nuclear power stations.

2.131 Several respondents thought that, because we have not considered the options for managing this issue adequately, we cannot conclude the ethical argument. A small number felt that the ethical question is a matter of principle which we can never fully resolve.

## Costs of waste management and decommissioning

2.132 A number of respondents felt it had not been made clear how waste disposal costs would be calculated. In particular they questioned the meaning of the reference to the "full share of costs" in the consultation document[152] and asked if it would be determined by volume or by level of radioactivity. Some felt that the private sector should be required to pay for the "full cost" of waste management and not a "full share". Others suggested that the owner/operator of a new nuclear power station should make all payments to cover waste and decommissioning costs into a fund at the time that the power station starts up. They suggested that, in view of uncertainties in economic performance over the very long-term, when calculating the payment to be made to the waste and decommissioning disposal fund we should use a zero discount rate when assessing any expenses to be incurred more than 30 years in the future. Some commented that investors in new nuclear power stations ought to pay only the cost of providing the additional space for the new waste in the geological disposal facility (the marginal cost) and that this cost should be made clear to investors. This cost could be levied as a charge per unit of electricity generated.

152 The Future of Nuclear Power, *The Role of Nuclear Power in a Low Carbon UK Economy, Consultation Document*, URN 07/970, May 2007.

2.133 Other respondents were not clear whether the present estimates of the cost of new nuclear power stations included the costs of waste and decommissioning. There was a view that this should be the case to enable proper comparison with other forms of electricity generation. Some expressed the view that the Government should not pay for decommissioning, while some questioned whether we should rely on energy companies to deal with the waste properly. They felt that the Government would need to put in place and apply very tight and transparent controls. There was also concern at any prospect of energy companies going bankrupt and leaving the clean-up costs to the taxpayer. This would require careful regulation and scrutiny by the Government. Given the long timescales involved, some respondents raised concerns about the level of certainty we could have in the costs. A small number of respondents felt that the Government should manage the process of fund accumulation itself to ensure that the costs of waste management are met.

## The Managing Radioactive Waste Safely (MRWS) programme

2.134 Respondents expressed a range of comments on an approach to siting a geological disposal facility based on voluntarism (that is, a willingness to participate)[153]. These comments varied from scepticism that any host community with a stable geology would volunteer, to agreement that the likely economic benefits of volunteering meant that communities and Local Authorities might put themselves forward. Some felt that communities would be more likely to agree to volunteer to host a geological disposal facility if it was just for legacy waste as there would be additional concerns around the disposal and storage of new nuclear waste. The MRWS consultation addressed the concept of voluntarism, and how it might relate to a volunteer community. Any material collected as part of the consultation on the future of nuclear power, which is relevant to the MRWS consultation, is being considered as part of that consultation. We have also analysed relevant responses to the consultation on MRWS and have factored them into our conclusions in this White Paper.

## *Government response*

2.135 The Government acknowledges the high degree of concern expressed about nuclear waste[154] at the consultation events and by some of the respondents to the consultation.

## Government policy on managing higher activity wastes

2.136 The Government is encouraged by the level of support for the view that a geological facility would be the best solution for managing both new and legacy waste. It is clear, however, that a significant number

153 The concept of voluntarism, and how it might relate to a volunteer community, was addressed in the Managing Radioactive Waste Safely consultation (Managing Radioactive Waste Safely, *A Framework for Implementing Geological Disposal*, 25 June 2007) which ran between 25 June and 2 November 2007.
154 Unless otherwise stated, references in this document to the Government position on waste refer to "higher activity waste", which includes intermediate level waste and spent fuel from any new nuclear power stations.

of respondents are concerned that the UK has yet to implement a long-term solution for existing waste.

2.137 We have made progress in developing a solution for waste management since 2003, by taking forward work through the MRWS programme to identify and set in place arrangements for the delivery of a safe, publicly acceptable solution for handling the UK's existing higher activity waste that would protect the environment and public safety.

2.138 As part of the MRWS programme, in 2003 the Government appointed CoRWM to assess and make recommendations on the best option or combination of options for the long-term management of the UK's higher activity wastes. Following an extensive UK-wide programme of engagement with the public and stakeholders, as well as with experts, CoRWM published its report in July 2006[155] and, in October 2006, the Government[156] accepted the recommendation that geological disposal, coupled with safe and secure interim storage, is the best available approach for the long-term management of existing higher-activity radioactive wastes and confirmed its support for exploring an approach based on voluntarism and partnership with local communities. CoRWM recognised that other management solutions may be appropriate for reactor decommissioning wastes because of the nature of the waste form. This could include interim decay storage or similar solutions to those arising from the new Low Level Waste (LLW) policy to allow the principal radiation emitters to decay. This is being taken forward as part of the MRWS programme.

2.139 The Government is taking forward its waste management policies formed in light of CoRWM's recommendations. This includes consulting on a phased approach to the process for implementing and siting a geological disposal facility, ensuring robust interim storage is available, continued research and development, and exploring how an approach based on voluntarism and partnership could be made to work in practice. The process began with a public consultation on implementation and the potential role of voluntarism in the process.

2.140 The Scottish Executive announced on 25 June 2007 that they did not endorse the decision by the UK Government and other Devolved Administrations to seek to develop a geological disposal facility through the Managing Radioactive Waste Safely (MRWS) programme. The stated position of the Scottish Executive is to support long-term "near surface near site" storage facilities, and they are have therefore disengaged from the MRWS consultation on a framework for implementing geological disposal. The UK Government will continue to work with the other Devolved Administrations (the Northern Ireland Assembly and the Welsh Assembly Government) to take forward the MRWS implementation framework. We will also continue to work with the Scottish Executive on all radioactive waste issues other than geological disposal.

155 Managing our Radioactive Waste Safely (CoRWM Document 700), *CoRWM's Recommendations to Government*, July 2006.
156 *Response to the Report and Recommendations from the Committee on Radioactive Waste Management (CoRWM).*

## Disposing of new build waste in a geological disposal facility

2.141 The Government has accepted CoRWM's recommendation that "Within the present state of knowledge geological disposal is the best available approach for the long-term management of all the material categorised as waste in the CoRWM inventory when compared with the risks associated with other methods of management[157]". The Government considers, based on scientific consensus and international experience, that despite some differences in characteristics, waste and spent fuel from new nuclear build would not raise such different technical issues compared with nuclear waste from legacy programmes as to require a different technical solution. The consultation[158] also considered these issues. The Government has thus concluded that it would be technically possible to dispose of waste from new nuclear power stations in a geological disposal facility. We have also considered whether there have been any further developments since CoRWM's recommendations including in response to the consultations[159] and Managing Radioactive Waste Safely[160] and we remain confident in the preliminary view we set out in the consultation that it would be technically possible to dispose of waste from new nuclear power stations in a geological disposal facility.

2.142 International experience backs up the belief that it would be technically possible to dispose of new waste in a geological disposal facility. In their final report to the Government, CoRWM stated "It became apparent early on that all countries with a nuclear power programme that have made decisions about long-term management of radioactive waste have adopted a strategy of interim storage followed by geological disposal"[161]. This includes countries such as Canada, Finland and the USA. We note that varying degrees of progress have been made towards delivering this objective in practice and that no geological disposal facility is yet operational for High Level Waste (HLW) or spent fuel. However, underground investigations are underway in Sweden and Finland into geological disposal facilities for spent fuel, following success in constructing geological facilities for Intermediate Level Waste (ILW) and LLW. In the USA a license application is being prepared to construct a geological disposal facility to dispose of HLW and spent fuel. Further detail can be found on page 129 of our consultation document[162]. The Government believes therefore that it is well-placed to benefit from international experience as the MRWS programme moves forward.

157 Managing our Radioactive Waste Safely (CoRWM Document 700), *CoRWM's Recommendations to Government*, July 2006. Full text of recommendation is "Within the present state of knowledge geological disposal is the best available approach for the long-term management of all the material categorised as waste in the CoRWM inventory when compared with the risks associated with other methods of management. The aim should be to progress to disposal as soon as practicable, consistent with developing and maintaining public and stakeholder confidence."
158 The Future of Nuclear Power, *The Role of Nuclear Power in a Low Carbon UK Economy, Consultation Document*, URN 07/970, May 2007.
159 The Future of Nuclear Power, *The Role of Nuclear Power in a Low Carbon UK Economy, Consultation Document*, URN 07/970, May 2007.
160 Managing Radioactive Waste Safely, *A Framework of Implementing Geological Disposal*, 25 June 2007.
161 Managing our Radioactive Waste Safely (CoRWM Document 700), *CoRWM's Recommendations to Government*, July 2006.
162 The Future of Nuclear Power, *The Role of Nuclear Power in a Low Carbon Economy, Consultation Document*, May 2007.

2.143 For completeness, we note that in its statement on nuclear new build, CoRWM stated that "solutions for existing and unavoidable future wastes would also be robust in the light of all reasonably foreseeable developments in nuclear energy and waste management practices"[163] although it felt that "significant practical issues would arise, including the size, number and location of waste management facilities "[164]. The Government acknowledges this and also that the focus of CoRWM's public and stakeholder engagement was always on the existing wastes and materials. The Government also acknowledges CoRWM's stated position that "its conclusions and recommendations are only intended to apply to committed wastes. It is important that CoRWM's views are not taken out of context"[165].

2.144 The Government and the NDA, as the organisation charged with responsibility for the programme to develop and deliver geological disposal, are committed to continuing research and development in radioactive waste management. Whilst the Government believes that geological disposal will provide a technically possible means of disposing of existing and new waste, the NDA will review alternative waste management options and, if deemed necessary, will undertake further research into those options[166]. This is in addition to the extensive programme of research that will be carried out during the development of the geological disposal programme as work progresses to assess a particular site or sites. All this will build on existing research and knowledge gained from the UK's work on radioactive waste disposal and experience gained overseas in geological disposal programmes.

2.145 Safety and environmental protection will be paramount in developing a geological disposal facility. The containment of radioactivity will be central to any safety case presented to the regulators. Unless the regulators can be satisfied that the risks that the radioactive contents pose to workers or the public can be made acceptably small, they would not permit the facility to be built and operated. The basic principle of a geological disposal facility is that it isolates the waste for so long and to such a degree that there is no significant surface exposure. In addition, the security of any facility will be strictly regulated by the OCNS. Given international experience and the UK's own research, we are confident that a geological disposal facility could be built in such a way as to satisfy the regulators. Safety, security and environmental protection will also be essential in ensuring that there is robust interim storage of waste before the geological disposal facility is developed, commissioned and available for use. Given the ability of interim stores to be maintained in order to hold waste safely and securely if necessary for very long periods (stores currently being constructed for the NDA are designed to last for at least 100 years), or if necessary refurbished or replaced, we are satisfied that it is reasonable to proceed with

163 CoRWM statement on Nuclear New Build 16 December 2005.
164 CoRWM statement on Nuclear New Build 16 December 2005, Addendum (March 2006).
165 Reiteration of CoRWM's Position on Nuclear New Build (CoRWM document 2162.2), September 2007.
166 Further detail is set out in the MRWS Consultation: Managing Radioactive Waste Safely, A Framework for Implementing Geological Disposal, 25 June 2007.

**91**

allowing operators to build new nuclear power stations in advance of a geological disposal facility being available.

2.146 In conclusion, having considered the evidence and arguments set out in the nuclear consultation document and the MRWS consultation and the responses to those consultations, the Government is satisfied that geological disposal would provide a technically possible means of disposing of higher activity wastes from new nuclear power stations.

## Impact of waste from new nuclear power stations on existing waste management strategy

2.147 The Government continues to believe that it would be technically possible and desirable to dispose of new waste in the same geological disposal facility as existing waste and that we should explore this through the MRWS process.

2.148 We note the number of responses in support of disposing of both new and legacy waste in the same geological disposal facilities although we recognise the concerns raised on this issue.

2.149 The Government also acknowledges that CoRWM considered that "should a new build programme be introduced...it would require a quite separate process to test and validate proposals for the management of wastes arising"[167]. The nuclear consultation document[168] set out the Government's views on the feasibility and desirability of disposing of new build waste in a geological disposal facility (or repository), including the balance of ethical considerations in relation to any decision to create new waste. The consultation also considered the impact that waste from new nuclear power stations would have on existing waste management strategies, including on the UK's waste inventory, and on the size and cost of a geological disposal facility. The nuclear power consultation process provided an opportunity to test those views.

2.150 We recognise the importance of being able to give as much clarity as possible to communities on the likely increases in both the volume and the level of radioactivity of the waste inventory that would arise from disposing of waste from new nuclear power stations in the same geological disposal facility as existing waste. The actual changes to the UK's waste inventory as a result of new nuclear power stations would depend on the number of stations that are constructed, among other factors. Further detail on the possible impact of new nuclear build on the existing waste inventory can be found on page 135 of the nuclear consultation document[169]. Through the MRWS programme, we will provide potential host communities with up to date information as the programme moves forward.

---

167 Managing our Radioactive Waste Safely (CoRWM Document 700), *CoRWM's Recommendations to Government*, July 2006.
168 The Future of Nuclear Power, *The Role of Nuclear Power in a Low Carbon UK Economy, Consultation Document*, URN 07/970, May 2007.
169 The Future of Nuclear Power, *The Role of Nuclear Power in a Low Carbon UK Economy, Consultation Document*, URN 07/970, May 2007.

2.151 In the nuclear consultation document[170], we set out that if new build waste were to be accommodated in the same geological disposal facility as legacy waste, this would affect the overall cost of the geological disposal solution as additional space would have to be provided and the design would need to be modified. It will be important to evaluate the final design of a geological disposal facility to accommodate legacy and new build waste, for operational and long-term safety. The fact that we have not begun construction of a repository at this time allows us to build in any necessary engineering features to accommodate particular types of waste if that proves necessary and publicly acceptable and the required safety case (including consideration of site location and geology) can be made. The size of any programme of new nuclear power stations will have an impact on whether all of the new waste could be stored in the same geological disposal facility as legacy waste. The Government proposes to pursue these issues through the MRWS programme.

## Identifying a suitable site for a geological disposal facility

2.152 The Government has confirmed its support for exploring how an approach to finding a site for a geological disposal facility based on voluntarism and partnership with local communities could be made to work. An approach based on voluntarism and partnership has been used in a number of countries, for example Finland and Sweden, as part of the process for the siting of geological disposal facilities for radioactive waste. Partnership is an assembly of local interests established to discuss, evaluate and advise on the potential implications of hosting a geological disposal facility. Overseas experience suggests that such an approach to local engagement is likely to be the most effective way of addressing the concerns and aspirations of communities considering hosting a geological disposal facility, whilst also providing a workable mechanism for identifying a suitable site.

2.153 The MRWS consultation considered the implementation of geological disposal and the potential role of voluntarism in the process. An analysis and summary of responses to that consultation has been published[171] and a White Paper is intended to follow in spring 2008 setting out the details of how the Government expects to see implementation taken forward. The Government believes that nothing has emerged from the MRWS consultation that alters our view on either geological disposal or on an approach based on voluntarism as a means of securing a site. A Government statement on the MRWS process and on geological disposal is set out in Box 1 and reproduced in Box 3.

2.154 We are aware that various sites for a geological disposal facility have been investigated to different degrees in the past, and that some concerns have been raised about the suitability of particular areas, although the suitability of individual sites would not be known until much more work has been done. It is important to maintain orderly progress towards a permanent solution for waste. The Government's

170 The Future of Nuclear Power, *The Role of Nuclear Power in a Low Carbon UK Economy, Consultation Document*, URN 07/970, May 2007.
171 http://www.defra.gov.uk/corporate/consult/radwaste-framework/index.htm

formal response to the MRWS consultation will be set out in the forthcoming White Paper and will focus on the issues around a geological disposal facility in more detail.

## BOX 3: GOVERNMENT STATEMENT ON THE MRWS PROCESS AND GEOLOGICAL DISPOSAL

- In October 2006, the Government[172] accepted the recommendation of the independent Committee on Radioactive Waste Management (CoRWM) that geological disposal was the best available approach to the long-term management of the UK's higher activity radioactive waste
- CoRWM's recommendations followed more than two and a half years' work assessing all of the available options on the basis of a wide programme of engagement with the expert community, stakeholder groups and the public
- CoRWM also recommended that progress towards geological disposal should be coupled with a robust programme of safe and secure interim storage. Again the Government accepted the Committee's recommendation saying that:

  "The design of new stores will allow for a period of interim storage of at least 100 years to cover uncertainties associated with the implementation of a geological repository. The replacement of stores will be avoided wherever possible, but the NDA will ensure that its strategy allows for a safe and secure storage of the waste contained within them for a period of at least 100 years."

- Delivery of these commitments by the Government and the Nuclear Decommissioning Authority (and its agents) will be supported by research and development programmes. Where appropriate, international programmes and experience will be drawn on. It is clear that geological disposal is the internationally preferred option for the long-term management of higher activity radioactive waste. There has been extensive progress towards delivery of geological disposal solutions internationally in recent decades. Within the next one or two decades, overseas geological disposal facilities are likely to become operational for spent fuel, in addition to the facilities that already exist for Intermediate Level Waste (ILW) and Low Level Waste (LLW).
- The Government also said in its response to CoRWM that it would explore the concept of voluntarism and partnership arrangements in delivery of geological disposal of the UK's higher activity radioactive waste. We set out proposals for doing this, and asked for people's views on the issue more widely in the June 2007 consultation document "Managing Radioactive Waste Safely: a Framework of Implementing Geological Disposal".

172 *Response to the Report and Recommendations from the Committee on Radioactive Waste Management (CoRWM).*

Department for Business, Enterprise and Regulatory Reform  MEETING THE ENERGY CHALLENGE

- This consultation closed on 2 November 2007. An analysis and summary of the responses has been published[173]. Overall there was general agreement with the Government's proposals, including that of seeking a voluntarism and partnership approach, although many detailed points were made.
- Following from CoRWM's recommendation (in relation to existing waste), international opinion and in line with the MRWS consultation, the Government continues to see geological disposal as the way forward for the long-term management of the UK's higher activity waste.

2.155 Given the international experience and the analysis of the responses to consultations on the future of nuclear power[174] and Managing Radioactive Waste Safely[175], we are satisfied that a geological disposal facility would provide a possible and desirable mechanism for disposing of new and legacy waste. We are also satisfied that there are feasible long-term mechanisms through the MRWS programme for identifying a suitable site and for constructing a geological disposal facility.

## Interim storage

2.156 A key part of CoRWM's recommendations was that "A robust programme of interim storage must play an integral part in the long-term management strategy"[176]. The MRWS consultation states:

"Existing stores for packaged waste are designed to provide a service life of 50 to 100 years or more. In the NDA's view these stores can have their service lives extended as required, in order to provide sufficient safe and secure interim storage throughout the geological disposal facility development programme. Subject to meeting regulatory safety and security requirements, any new interim stores on NDA sites will have service lives of 100 years or more"[177].

2.157 CoRWM extensively examined the options available for the disposal of existing radioactive waste, which included interim storage prior to geological disposal. As part of their analysis, CoRWM examined evidence from other countries and concluded that "Knowledge of international experience has contributed greatly to CoRWM's work. It has shown that, outside the UK, all countries with a nuclear power programme have selected interim storage followed by geological disposal as their strategy for managing long-lived waste"[178].

173 http://www.defra.gov.uk/corporate/consult/radwaste-framework/index.htm
174 The Future of Nuclear Power, *The Role of Nuclear Power in a Low Carbon UK Economy, Consultation Document*, URN 07/970, May 2007.
175 Further detail is set out in the MRWS Consultation: Managing Radioactive Waste Safely, *A Framework for Implementing Geological Disposal*, 25 June 2007.
176 Managing our Radioactive Waste Safely (CoRWM Document 700), *CoRWM's Recommendations to Government*, July 2006.
177 Managing Radioactive Waste Safely, *A Framework of Implementing Geological Disposal*, 25 June 2007.
178 Managing our Radioactive Waste Safely (CoRWM Document 700), *CoRWM's Recommendations to Government*, July 2006.

2.158 To implement the Government's policy on interim storage, the NDA is already undertaking a UK-wide review of existing waste storage facilities. Once this is complete, the NDA will consider what may additionally be required to fulfil the Government's commitment to ensure robust interim storage until there is a final disposal solution. The NDA is committed to ensuring that safe and secure interim storage will be available for its expected waste until a geological disposal facility has been constructed. For new build waste the provision of such stores would be the responsibility of the operator of the nuclear power station. Security of radioactive waste storage and transport is kept under constant review by the regulators to ensure that facilities and practices remain robust. The NDA will continue to work with the environmental, safety and security regulators to ensure that they are all satisfied that these facilities meet their strict requirements.

2.159 From the conclusions drawn by CoRWM in relation to existing waste and the advice of the NDA, we are satisfied that interim storage will provide an extendable, safe and secure means of containing waste for as long as it takes to site and construct a geological disposal facility. The nuclear consultation[179] also considered this issue. Section 3 of this White Paper sets out that operators of new nuclear power stations will be required to pay for, and ensure the availability of, interim storage for waste until we expect a geological disposal facility to be in a position to accept waste from new nuclear power stations, and beyond that date to provide adequate contingency.

## Managing Low-Level Waste (LLW)

2.160 On 26 March 2007 the Government announced an update of its policy for LLW management[180]. Under the new policy, the NDA is now responsible for developing and maintaining a national strategy for handling LLW from nuclear sites and for ensuring continued provision of the waste management and disposal facilities required. The LLW strategy that the NDA develops will be reflected in its annual plans and strategy document in due course which will be subject to public consultation.

## Decommissioning new nuclear power stations

2.161 The consultation document describes the existing UK decommissioning strategy and describes various issues in relation to the decommissioning of new nuclear power stations. The Government acknowledges comments made in response to the consultation on the decommissioning of nuclear power stations. We agree with those who pointed out that decommissioning of modern reactors is expected to be considerably easier than the work the NDA is currently doing to decommission our existing nuclear facilities. We also agree that it will be important to ensure that lessons can be learned from international experience of decommissioning.

179 The Future of Nuclear Power, *The Role of Nuclear Power in a Low Carbon UK Economy, Consultation Document*, URN 07/970, May 2007.
180 http://www.defra.gov.uk/environment/radioactivity/waste

## Ensuring that operators meet their full decommissioning costs and their full share of waste management costs

2.162 We note and agree with the comments made by some respondents that operators of new nuclear power stations need to make adequate and secure financial provision to meet the full costs of decommissioning and their full share of waste management and disposal costs. Through the Energy Bill, the Government will put in place robust arrangements to ensure that operators set aside sufficient funds in a secure way to cover their full decommissioning costs and their full share of waste management costs. Section 3 of this White Paper describes our proposed approach for achieving this. We can confirm that, to ensure a fair comparison between the costs of nuclear and other types of generation, our economic modelling includes an estimate of the costs of decommissioning and waste management, although further work will be done in the coming years to refine these numbers (further detail can be found in the Government's response to Question 4 of the consultation).

## The ethical considerations of allowing new nuclear waste to be produced

2.163 The Government agrees that the creation of new waste raises ethical issues. The consultation provided an opportunity for respondents to raise and provide their considered views on these issues. The Government has considered the comments that nuclear waste would be an unacceptable legacy for us to leave to future generations. We have also noted the arguments put forward that nuclear power may provide significant benefits to future generations, particularly in terms of reducing $CO_2$ emissions, as well as improving security of energy supply, which will help to ensure that future generations have access to the same or a better standard of living to the one that we currently enjoy.

2.164 The Government believes that the intergenerational issues of radioactive waste should not be considered in isolation, but alongside the long-term impact of climate change. If no new nuclear power stations are built there would be no additional radioactive waste. However, there could be negative consequences for the environment, due to increased $CO_2$ emissions if some fossil fuel power stations, without CCS technology, are built to meet energy demand instead of nuclear. The economic modelling set out in the consultation document[181] suggests that without the option of nuclear power, this would be a likely scenario in the medium term (up to 2030), because coal and gas fired power stations remain more economic compared to renewables, and CCS is a new technology which is not currently operating on a commercial scale.

2.165 Renewables are another option and the Government is committed to ensuring that renewables will make an increasing contribution to energy supply in the UK. However, if all the existing nuclear capacity were replaced by wind power alone, it would take 25 GW[182] of wind

---

181 Chapter 5, The Future of Nuclear Power, *The Role of Nuclear Power in a Low Carbon UK Economy, Consultation Document*, URN 07/970, May 2007.

182 This is because of the intermittency of wind power. The turbines only produce electricity when local weather conditions are favourable.

capacity, when currently around 2 GW is available. Assuming a turbine size of 2 MW, this would mean more than 12,000 turbines. Each GW of wind power would cover around 10,000 hectares of land[183]. In addition, as wind is an intermittent type of generation, and is not available at all times, it would not provide a reliable baseload supply of energy. Besides nuclear, the only other proven low-carbon form of generation of baseload electricity is large-scale hydro, which as a technology has limited potential for further capacity increases in this country as many suitable sites have already been exploited. Without new nuclear, and considering the uncertainty surrounding the development of CCS, it seems likely that a significant proportion of the new capacity built to meet baseload demand for energy will come from additional fossil fuel power stations.

2.166 We have also considered the extent to which the burdens to future generations can be mitigated. Radioactive waste and $CO_2$ emissions are both potentially hazardous and will impact on both current and future generations. Our understanding of radioactive waste and how to deal with it is arguably more advanced than our knowledge of the impact of man-made climate change and as yet we have no solution for mitigating the risks posed by increased $CO_2$ emissions. We have no solution for reversing the adverse global environmental effects of these emissions, whereas we believe that geological disposal will provide a technically possible mechanism for disposing of radioactive waste safely that is already being taken forward in several other countries.

2.167 On balance, we believe that not taking action now on climate change raises more significant inter-generational challenges than does the management of radioactive waste. By putting our existing wastes into passively safe forms, building interim storage and working towards the development of geological disposal facilities for waste disposal, we are making progress in dealing safely with radioactive waste for the long-term now rather than leaving the problem for future generations, whereas we have no equivalent way of mitigating the potential impacts of climate change other than by increasing our use of low-carbon forms of generation. The Government is committed to reducing $CO_2$ emissions by all means possible. We believe that nuclear power can play a part here, alongside other means, such as better energy efficiency and research and investment into other forms of low-carbon generation.

2.168 We accept that a geological disposal facility cannot be completed for some decades. We have considered carefully whether it is right to proceed with allowing new nuclear power stations to be built before a geological disposal facility is constructed. In practice, as this will be many years in the future it would rule out the ability of nuclear power to contribute to the new capacity required in the next twenty years as existing nuclear and fossil fuelled stations reach the end of their lives. Given the progress being made in developing and implementing policy for long-term waste management through the MRWS and CoRWM processes, the commitment to carrying forwards a geological disposal programme and the ability of interim stores to hold waste safely and

183 British Wind Energy Association, http://www.bwea.com/ref/faq.html.

securely if necessary for very long periods, we are satisfied that it is reasonable to proceed with a nuclear programme in advance of a geological disposal facility being available.

## *Our conclusion*

**Having reviewed the arguments and evidence put forward, the Government believes that it is technically possible to dispose of new higher-activity radioactive waste in a geological disposal facility and that this would be a viable solution and the right approach for managing waste from any new nuclear power stations. The Government considers that it would be technically possible and desirable to dispose of both new and legacy waste in the same geological disposal facilities and that this should be explored through the Managing Radioactive Waste Safely programme. The Government considers that waste can and should be stored in safe and secure interim storage facilities until a geological facility becomes available.**

**Our policy is that before development consents for new nuclear power stations are granted, the Government will need to be satisfied that effective arrangements exist or will exist to manage and dispose of the waste they will produce.**

**The Government also believes that the balance of ethical considerations does not rule out the option of new nuclear power stations.**

# Nuclear power and the environment

**THE GOVERNMENT'S PRELIMINARY VIEW**

The Government believes that the environmental impacts of new nuclear power stations would not be significantly different to other forms of electricity generation and given the UK and European requirements in place to assess and mitigate the impacts, that they are manageable. Therefore, the Government believes that they do not provide a reason to not allow energy companies the option of investing in new nuclear power stations.

We recognise the need for a strategic assessment of the environmental issues relating to new nuclear power stations. If the Government confirms its preliminary view that it is in the public interest to allow energy companies the option of investing in new nuclear power stations, we propose to undertake an SEA as part of a Strategic Siting Assessment, the detail of and proposed timetable for which were set out in a detailed consultation alongside our nuclear consultation on the issue in principle.

**Question 11**
**Do you agree or disagree with the Government's views on environmental issues? What are your reasons? Are there any significant considerations that you believe are missing? If so, what are they?**

## The key arguments and issues presented in responses

2.169 Respondents expressed differing views on nuclear power and the environment. These both supported and opposed the Government's position. Some respondents were concerned about the potential environmental impact of a nuclear accident, which they considered to be much greater than the impact of an accident at a wind farm. In addition, in their comments to this question, a large proportion of those who opposed the Government's view considered nuclear waste to be an overriding issue from an environmental perspective. Some respondents also expressed concerns about the environmental impacts of discharges from the normal operation of a nuclear power station.

2.170 A number of respondents said that the consultation document underplayed environmental aspects of uranium mining, which is unique to nuclear power among electricity generation technologies. Others also pointed to the potential need to put in place new grid infrastructure if any nuclear stations are sited in remote locations where there are currently no links to the grid.

2.171 Some respondents highlighted the fact that nuclear power requires much smaller quantities of fuel than needed for fossil-fuel generation, which has environmental benefits in terms of less mining and transportation. Others commented that the environmental impacts of nuclear power stations need to be addressed at a national level, welcoming the proposal for a Strategic Environmental Assessment

(SEA). Others said that the environmental impacts of any new nuclear power stations would be more manageable if they were confined to existing sites.

2.172 A key concern of some respondents relates to the proposed reforms to the planning system as set out in the Planning White Paper. The concern is that the changes to the planning system could remove the rights of local people in decisions on key infrastructure projects, such as new nuclear power stations.

2.173 Some respondents took issue with the claim in the consultation document that the impact of a nuclear power station on the landscape is comparable to that of a wind farm. Some said that the impact of a wind farm is much less than that of a nuclear power station. The main reasons cited for this were:
- the land taken by a wind farm can be re-used as soon as the turbines are taken down, which is not the case with a nuclear power station
- wind farms do not cover over most of the land in which they are situated – most of the land taken can still be used for grazing, growing crops, wildlife habitation etc
- wind turbines can be built off-shore.

2.174 Other respondents said the environmental impact of wind farms is actually much greater than that of nuclear power stations. The main reasons cited for this were:
- wind farms are often sited in remote areas and therefore need a great deal of new grid infrastructure
- wind farms make more noise and have a greater and more negative visual impact
- wind farms require costly back-up generators in periods of no wind.

2.175 Wind farms aside, respondents generally accepted the view presented in the consultation document, namely that the landscape impacts of nuclear power stations do not differ significantly from other forms of electricity generation.

## Government response

2.176 We acknowledge the legitimate concerns that people have expressed over the potential for serious environmental damage in the event of a major accident at a nuclear power station. As we have already explained (paragraph 2.87), the UK has a strict nuclear safety regime, enforced by an independent, nuclear inspectorate (HSE's NII). Nuclear power stations are required to be designed to cope with a wide range of potential failures of equipment or of operation, in accordance with HSE's Safety Assessment Principles. They must also satisfy stringent environmental requirements enforced by the relevant environmental agency. Furthermore, some new designs of nuclear reactors have "passive" safety features which rely on natural processes to shut down the plant safely in the event of serious operational problems, which are designed to make them more robust against major accidents leading to

**101**

a release of radioactivity. Current reactors, such as the AGR stations in the UK, and any modern reactors that might be built in this country, are designed to ensure that major accidents leading to a significant release of radioactivity have very low probability of occurrence.

2.177 With specific regard to the environmental risks arising from dealing with waste from any new nuclear power station, as we have explained (paragraphs 2.147-2.151), we believe that it would be desirable and technically possible to dispose of legacy and new waste safely in the same geological disposal facility, and store in safe and secure interim storage until such a facility becomes available. With regard to the concern raised by some respondents on discharges during normal operation, nuclear power stations must satisfy stringent environmental regulations enforced by the relevant environment agency. The environment agencies require that radioactive waste created and discharges made are minimised. The UK remains committed to meeting its obligations under the OSPAR Convention[184] on the protection of the marine environment of the north-east Atlantic, in respect of discharges of radioactive substances. The UK Strategy for Radioactive Discharges 2001-2020 (currently being revised) provides the framework for achieving this.

2.178 The Government accepts that we cannot ignore the environmental impact of uranium mining. The waste created through uranium mining can vary greatly, depending on the concentration of uranium and how it is mined. An underground mine may generate less than one tonne of waste rock for every tonne of ore produced. In comparison, in extreme cases an open-pit mine may generate 40 tonnes of waste rock for every tonne of uranium ore. As we indicated in the consultation document, however, conventional uranium mining does not differ significantly from mining of other metalliferous ores or coal for other types of power station. Furthermore, an increasing proportion of the world's uranium now comes from in-situ leaching. As we acknowledged in the consultation document, this is a process that does not require the ore to be mined and generates much less waste, though it can have a negative impact on the water table and is not suitable for all types of uranium deposits. There are established environmental constraints, such as the regulations governing uranium mining in Australia which cover, amongst other things, environmental protection and the requirement to meet environmental approvals before mining proceeds. Additionally, most uranium mining companies in Australia and Canada, which supply much of the world's uranium, have achieved certification from the International Organisation for Standardisation. This body sets the standard for, and undertakes audits of, environmental management systems. These environmental constraints minimise the environmental impacts of mining operations.

2.179 The Government agrees that we need to address the high-level environmental impacts of any new nuclear power stations at national level, without removing the need for site-specific environmental assessments. As we explain (see Section 3) we propose that as part

184 The Oslo Paris Convention, 1992.

of the proposed Strategic Siting Assessment (SSA) a SEA should be carried out to examine the environmental impacts of nuclear power.

2.180 We have also considered the concerns, echoing comments received through the consultation on the Planning White Paper, that the proposed changes to the planning system will remove the rights of local people in decisions on nationally significant infrastructure projects. In taking forward the proposals set out in the Planning Bill, currently before Parliament, the Government will ensure engagement and consultation with communities affected by any planning proposal. Section 3 of this White Paper discusses in detail these and other proposed changes to the planning system.

## *Our conclusion*

**Having reviewed the arguments and evidence put forward, the Government believes that (with the exception of the waste issue discussed above) the environmental impacts of new nuclear power stations would not be significantly different to those of other forms of electricity generation and that they are manageable, given the requirements in place in the UK and Europe to assess and mitigate the impacts. Therefore, the Government believes that environmental impacts do not provide a reason not to allow energy companies the option of investing in new nuclear power stations.**

**In confirming the Government's view that it is in the public interest to allow energy companies the option of investing in new nuclear power stations, we propose to undertake a Strategic Environmental Assessment as part of a Strategic Siting Assessment.**

**103**

# The supply of nuclear fuel

**THE GOVERNMENT'S PRELIMINARY VIEW**

Based on the significant evidence that there are sufficient high-grade uranium ores available to meet future global demands, and the relatively small impact that allowing energy companies to invest in new nuclear power stations in the UK would have on global demand for uranium, the Government believes that there should be sufficient reserves to fuel any new nuclear power stations constructed in the UK.

**Question 12**
**Do you agree or disagree with the Government's views on the supply of nuclear fuel? What are your reasons? Are there any significant considerations that you believe are missing? If so, what are they?**

## Key arguments and issues presented in responses

2.181 Those who supported the Government's view on supply of uranium fuel agreed that there are sufficient high-grade uranium ores to meet future global demands. However, a number of people were concerned that a shortfall in the supply of uranium could occur in the near future, and that this might compromise energy security.

2.182 To counter any potential shortfall in uranium supply, the following suggestions were made:
   - utilities in the UK should stockpile fuel
   - we should investigate the use of thorium as a fuel
   - we should consider reprocessing (recycling) to extend the use of nuclear fuel
   - the use of fast breeder reactors potentially to extend uranium fuel supplies should be considered
   - the UK should explore the mining of uranium ore in the UK to reduce uranium imports
   - we should think about using the UK's existing stocks of uranium and plutonium as fuel
   - the use of uranium from "low-grade"[185] ores may be necessary because "high-grade"[186] ores may soon become exhausted

## Government response

2.183 The most recent Euratom Supply Agency (ESA) Report 2006[187] supports the Government's view that uranium resources will be available for the period likely to be covered by the operation of new nuclear power

---

185 Low-grade ore is defined by some authors as ore containing a maximum of 0.01% $U_3O_8$, for example, *CO$_2$ Emissions from the Nuclear Fuel Cycle*, M. Diesendorf & P. Christoff, November 2006.
186 High-grade ore is defined by some authors as ore containing a minimum of 0.1% $U_3O_8$, for example, *CO$_2$ Emissions from the Nuclear Fuel Cycle*, M. Diesendorf & P. Christoff, November 2006.
187 Euratom Supply Agency, *Annual Report 2006*.

stations in the UK and that more focused exploration will lead to increased availability over time. The recent Fourth Assessment Report of Working Group III of the Intergovernmental Panel on Climate Change provides further evidence when it states "Even if the nuclear industry expands significantly, sufficient fuel is available for centuries."[188]

2.184 Since the start of the UK civil nuclear industry we have depended on imported uranium to fuel our nuclear power stations. As explained in paragraphs 2.37-2.38, we do not believe that uranium resources or the future price of uranium will be limiting factors for a new nuclear build programme given that the cost of uranium is a relatively small component of the cost of nuclear generation. Uranium is supplied from a wide range of countries and we fully endorse the ESA recommendations[189], to the effect that utilities should cover most of their needs under long-term contracts with diversified supply sources. The Government also supports the action recommended by ESA for operators to maintain a sufficient stockpile of fuel to mitigate against potential supply interruptions.

2.185 The Government is aware of the potential use of thorium as a fuel for nuclear generation. Several countries, including the UK, have carried out research into using thorium as a nuclear fuel[190]. However, there are currently no commercially available reactors which utilise a thorium fuel cycle[191]. Industry proposals for a commercial thorium fuelled reactor would therefore appear unlikely to come forward while there are sufficient supplies of uranium available.

2.186 We deal with reprocessing at paragraphs 2.217-2.227. In principle, reprocessing, and particularly reprocessing together with the use of fast breeder reactors, could significantly extend the amount of energy which can be extracted from uranium. However, this is not the subject of our consultation, and is not considered further here.

2.187 Uranium is not mined in the UK. The British Geological Survey[192] (BGS) has identified several areas in the UK where uranium deposits exist, however, these are considered uneconomic to recover.

2.188 The NDA owns around 51,000 tonnes of uranium and 86.5 tonnes of plutonium[193]. The NDA is considering various options for dealing with this inventory including future fuel use as Mixed Oxide Fuel (MOX)[194]. The NDA have estimated that these materials could be converted into sufficient fuel to power up to three modern 1000 MW Pressurised Water Reactors (PWRs) for around 60 years. The evaluation for the future of the UK's plutonium stocks will take account of advice such as

188 R.E.H. Sims, et al 2007: Energy supply. In Climate Change 2007: Mitigation. Contribution of Working Group III to the Fourth Assessment Report of the Intergovernmental Panel on Climate Change [B. Metz, et al (eds)], Cambridge University Press, Cambridge, United Kingdom and New York, NY, USA.
189 Euratom Supply Agency, *Annual Report 2006*.
190 *Thorium Fuel for Nuclear Energy – Now You're Cooking with Thorium*, American Scientist, Volume 91, No. 5, p 408, October 2003.
191 World Nuclear Association, *Thorium*, September 2007.
192 British Geological Survey Mineral Profile, *Uranium*, March 2007.
193 NDA's response to The Future of Nuclear Power consultation, 5 October 2007.
194 MOX fuel consists of plutonium oxide that is blended with depleted uranium, left over from an enrichment plant, to form fresh mixed oxide fuel (MOX, which is $UO_2+PuO_2$).

The Royal Society's report[195] on the Strategy for the UK's Separated Plutonium which recommended the use of the UK's plutonium stocks as MOX fuel.

2.189 The Government has seen no evidence that "high-grade" ores are close to exhaustion. However, we have seen evidence that as prospecting for new uranium resources increases there are very encouraging "high-grade" deposits being found at the initial investigation stages, for example in Zambia[196] and Sweden[197], which are representative of similar deposits in several other countries. As explained as paragraph 2.37, nuclear fuel supply is a stable and mature industry.

### Our conclusion

**Having reviewed the arguments and information put forward, and based on the significant evidence that there are sufficient high-grade uranium ores available to meet future global demand, and the relatively small impact that allowing energy companies to invest in new nuclear power stations in the UK would have on global demand for uranium, the Government believes that there should be sufficient reserves to fuel any new nuclear power stations constructed in the UK.**

195 The Royal Society, *Strategy options for the UK's separated plutonium*, 21 September 2007.
196 Equinox Minerals Limited, *Lumwana Uranium Feasibility Study at Malundwe Delivers High Grade Uranium Intercepts*, Press release, 24 July 2007.
197 Mawson Resources Ltd., *Mawson Identifies High-Grade Uranium at Tresjöarna in Sweden*, News release, 23 April 2007.

# Supply chain and skills capacity

**THE GOVERNMENT'S PRELIMINARY VIEW**

The Government believes that the international supply chain and skills market should be able to respond if the Government were to allow energy companies to invest in new nuclear power stations. This view is based on:

- the long lead times associated with new nuclear power stations;
- the financial incentives for the private sector to meet the demands created by the building of new nuclear power stations; and
- the facilitative work that Government, the academic sector and industry are undertaking to support skills development in the relevant sectors.

Therefore, the Government believes that the supply of skills and supply chain capacity do not provide a reason to prevent energy companies from investing in new nuclear power stations.

**Question 13**
**Do you agree or disagree with the Government's views on the supply chain and skills capacity? What are your reasons? Are there any significant considerations that you believe are missing? If so, what are they?**

## *Key arguments and issues presented in responses*

2.190 The responses to the question of supply chain and skills capacity were varied but a few clear themes emerged. The most common response agreed with the Government's view that the supply chain and skills market present challenges but that these are manageable and do not provide a reason for ruling out new nuclear build. A key qualification that some respondents voiced was the need to move quickly to ensure that we retain essential skills and transfer them to a new generation and also to reverse the decline in the UK's ability to manufacture essential components.

2.191 Some respondents expressed the view that the situation is more challenging and that the UK needs to invest urgently in education, training, and skills development. Many of this group added that new nuclear build should take place as soon as possible, so that the UK can make use of the existing, but ageing, pool of skills and can secure a position at the front of the queue for technology and manufactured equipment. Some suggested that the Government should influence the timing of any investment, rather than allowing developers to decide, based on their reading of the market situation.

2.192 Some respondents questioned the skills and experience of labour sourced outside the UK and raised security implications of using such labour on nuclear sites.

**107**

2.193 Some respondents expressed the view that the UK's manufacturing and skills base has deteriorated to the point where it is no longer feasible to build new nuclear power stations in the UK. Some questioned the availability of manufactured equipment from foreign sources.

2.194 Others also argued that we would do better to spend resources developing other forms of power. Almost all of these respondents made such comments as part of a general objection to nuclear power.

2.195 In addition, there was a range of more general comments, many not strictly related to the issues of supply chain and skills. Of these, the most relevant referred to the need to have a firm programme for building nuclear power stations, so that investment in training and manufacturing capacity could be made with more confidence.

## Government response

2.196 We recognise that a programme of new nuclear power stations would have to progress on a realistic timescale, if we are to utilise and transfer existing skills before they are lost and if the supply chain is to be managed effectively. The exact timing is, of course, a judgement for energy companies, but there are key actions for Government to undertake to reduce the uncertainties in the pre-construction period through improvements to the regulatory and planning processes. This package of measures was set out in the nuclear consultation document[198] and is set out in Section 3 of this White Paper. These measures should increase investor confidence and encourage the market to invest in training and manufacturing.

### The challenge faced by the supply chain

2.197 The Government accepts that the supply chain for key components for new nuclear power stations will present challenges. Globally, over 1000 GW of old fossil fuel plant needs to be replaced or thoroughly upgraded over the next 25 years, as well as an ageing nuclear fleet. The developing world will need at least another 1000 GW of new capacity to support economic growth[199]. The resulting demand for equipment is likely to exceed the world's manufacturing capacity, at least some of the time. Engineering and construction services could also be severely stretched. Careful management of the supply chain will be essential to minimise delay and cost escalation, regardless of what type of plant is being built. While this might be a concern for project developers, it represents a major opportunity for the supply chain, especially manufacturers with the capability to enter, or re-enter, this market.

2.198 Manufacturing for the nuclear industry has declined in the UK but some important capacity remains, or is capable of recovery. For example,

---

198 The Future of Nuclear Power, *The Role of Nuclear Power in a Low Carbon UK Economy, Consultation Document*, URN 07/970, May 2007.
199 From sources including the IEA (http://www.worldenergyoutlook.org/) and the US department of Energy (http://www.eia.doe.gov/oiaf/ieo/electricity.html).

manufacturers in the UK could produce key components of the nuclear primary circuit – that is the pressurised water circuit – including the pipework, pressuriser, steam generators and pumps, as well as the containment structure[200]. If manufacturers in the UK decide not to pursue this business, there are established manufacturers elsewhere. Some of these are investing in extra capacity, although pinch points remain. Only the reactor vessel and its head closure would have to be made overseas (as they were for Sizewell B). It would take significant investment to develop a capability to manufacture reactor vessels in the UK and a decision to do this would depend on how companies view the global, rather than just UK, market opportunities.

2.199 The non-nuclear elements of the supply chain present a mixed picture. The UK no longer manufactures turbine-alternators for new power stations, so these would have to be imported. Transformers can be made in the UK, as can switchgear. All of these manufacturers have long order books at times of high demand. Civil engineering, structural steelwork and balance of plant are all within the UK capability, although nuclear power stations would have to compete with other projects, as explained later in this Section.

2.200 The biggest non-labour cost of a nuclear power station is the concrete, reinforcing bar, structural steel, pipe and cable that are bought in large quantities. At the time of writing, high demand for bulk supplies in China and the Middle East is impacting on price and availability worldwide. Domestically, there is greater potential for competition from other UK construction projects. It is therefore as important to manage the supply of bulk materials as it is the supply of big components, such as turbines, if costs are to be controlled effectively.

2.201 It is important to remember that all manufacturers, UK or overseas, participate in a global market. They sell worldwide and order books get longer if demand is high. If new manufacturing capacity is created in the UK, it will improve the supply situation overall, but project developers will still have to manage the situation to ensure the timely delivery of equipment.

## An ageing workforce

2.202 Across the energy sector in the UK, large numbers of workers will leave for retirement in the next decade[201]. Ensuring a continuity of skills and experience will be a challenge for human resource management. New nuclear build is challenging but, if nuclear power stations are not built, alternative capacity will be needed anyway and this will face similar resource pressures. For example, clean coal with CCS directly competes for process specialists with the oil and gas, refining and petrochemical industries, where there is exposure to the global market, with high demand for skills from overseas, especially the Middle East.

---

200 See reports by the Nuclear Industry Association at http://www.niauk.org/position-papers.html and IBM Business Consulting Services, *An Evaluation of the Capability and Capacity of the UK and Global Supply Chains to Support a New Nuclear Build Programme in the UK*, IBM UK Ltd, Basingstoke, 2005.

201 There is extensive information on this subject from Cogent (http://www.cogent-ssc.com), Energy & Utility Skills (http://www.euskills.co.uk/) and the Engineering Construction Industry Training Board (http://www.ecitb.org.uk/).

**109**

2.203 The age profile in the UK's nuclear workforce follows the general trend, albeit with some degree of variation across the energy sector[202]. There is time to ensure a skills succession but this must start soon. An immediate concern for nuclear-specific skills is the design and licensing of the reactor, where skills across the board, from reactor physicists to safety case specialists, are both ageing and in short supply, here and around the world. This is the most challenging issue in the short-to-medium-term. It requires careful management of resources plus a large increase in university output of engineers and scientists, especially from specialised Masters-level courses. The Government is working through the Sector Skills Councils (SSCs) and the Engineering and Physical Sciences Research Council (EPSRC) to deliver a coherent skills strategy in this area.

## Competition from other projects

2.204 In engineering construction more generally, activity in the UK and worldwide is growing and new nuclear build would face competition from other projects. Construction projects, such as for the 2012 Olympics and Thameslink, have some impact on specialised skills, for example steel erectors and project management. While these should largely be complete by the time the UK starts to build new nuclear power stations, other projects, such as Crossrail and Thames Gateway, will come along so the UK will need a resource for these projects into the long term. Other energy, petrochemical and pharmaceutical sector projects compete more directly for core engineering construction skills, as do Ministry of Defence projects, such as new aircraft carriers. Supply will fall behind demand at times and many energy sector construction projects will face competition for skills, which will have to be managed. If investors decide on a planned fleet build of identical power stations, whether nuclear or not, that would be easier to manage than one-off projects.

## The global resource

2.205 Apart from some specialist jobs in the nuclear part of the station, technically known as the nuclear island, most of the skills and resources needed to build new nuclear power stations are generic to large engineering construction projects and have a wider source of supply. Given the global situation, we can expect the market to respond by delivering new capacity to build generating plant of all types from coal, to nuclear, and to renewables.

## Will nuclear build compete with renewables?

2.206 As we have already said elsewhere in this White Paper, the Government is committed to increasing the share of electricity from renewable technologies through the Renewables Obligation and the Climate Change Levy. The Renewables Obligation requires electricity suppliers to obtain an increasing proportion of their electricity from renewable sources or to pay a penalty for any shortfall.

202 From recent analysis undertaken by the NDA (http://www.nda.gov.uk/), Cogent (http://www.cogent-ssc.com) and the National Skills Academy for Nuclear (http://www.nuclear.nsacademy.co.uk/).

2.207 Renewables and nuclear power are key components in the Government's strategy for meeting the 2050 target for carbon dioxide emissions reduction, along with other low-carbon generating technologies, energy efficiency and demand reduction. Nuclear power requires specialised skills, both for construction and operation, but in modest numbers compared to the overall energy workforce. The rest of the skill sets are generic to engineering construction and can be deployed across a range of projects. Big projects like power stations have some skills overlap with large-scale renewables, such as wind or tidal barrage, especially for electrical equipment. Small-scale renewables are largely dependent on general building trades, where there is little skills overlap with nuclear power. Moreover, renewables technology is moving towards smart networks, which require electronics and IT skills, not those associated with heavy construction. There is no evidence to suggest that building new nuclear power stations would add extra pressure on the supply of skills to the renewables sector. In fact, it may well encourage a renaissance in science and engineering, benefiting the entire energy sector.

## What is being done about skills in the nuclear industry?

2.208 The Energy White Paper[203] asked the SSCs to report on the situation across the energy sector, including details of skills shortages, skills gaps (that is workers without all the skills required for their job), and the impact of demographic factors. This will include a forward look that takes account of factors such as retirement and new investment. It will set out the strategies the SSCs and employers are implementing to ensure that the UK can meet future skills needs. It will also consider the actions that can be taken to coordinate recruitment and training and mitigate damaging competition in the labour market. Government is working with the SSCs to deliver this report in the first half of 2008.

2.209 Early in this decade, the nuclear industry undertook a strategic review of its skills base, its future needs for skills, and the impact of the workforce demographics[204]. When the SSC, Cogent, took responsibility for the nuclear sector in 2004, it was able to build on this in developing its Sector Skills Agreement[205]. This is the industry-wide plan that sets out the strategy for future skills development, taking account of both the age profile and the skills gaps that are increasing, as workers are re-deployed from operations to decommissioning. The Sector Skills Agreement sets out a detailed analysis and an action plan to ensure that the industry's skills needs are met.

2.210 Cogent, with support from the NDA and a range of employers from the industry, recognised that a National Skills Academy for Nuclear (NSAN) could play a significant part in recruiting and developing the right skills[206]. Therefore, they submitted a bid in Round 2 of the NSA selection process and were invited in October 2006 to move into the business planning stage. A team seconded from the North West

203 Energy White Paper, *Meeting the Energy Challenge*, URN 07/1006, May 2007.
204 Nuclear and Radiological Skills Study, *Report of the Nuclear Skills Group*, 5 December 2002.
205 Available on http://www.cogent-ssc.com.
206 http://www.nuclear.nsacademy.co.uk/

Development Agency, together with a shadow board from employers has subsequently developed a detailed business plan. This was appraised by the Learning and Skills Council and the approval of NSAN as a National Skills Academy was announced by David Lammy, Skills Minister, in September 2007. The Academy will be formally launched early in 2008.

2.211 NSAN will build on and coordinate existing training provision on a national and regional basis to ensure it is aligned with employers' requirements and, with its training partners, aims to deliver 1000 apprenticeships, 150 foundation degrees and to re-train 4000 existing workers in its first three years of operation. This will be as part of a coherent skills strategy that will address the decommissioning of existing facilities, the on-going needs of the power generation industry, the Royal Navy propulsion programme, and new nuclear build if required. NSAN will also develop a strategic approach to higher education better to integrate technician and graduate training and to improve the supply of graduates into the sector.

2.212 In parallel, the Nuclear Employer Skills Group (NESG), formed of employers, Government Departments and Cogent, has been taking forward important work on career pathways, up-skilling, competence assurance, passports, qualifications credit frameworks, project management and foundation degrees.

2.213 The main Government agency for funding research and training in engineering and the physical sciences, the Engineering and Physical Sciences Research Council (EPSRC) is contributing £1 million and industry partners £1.6 million towards a "Nuclear Technology Education Consortium" to provide masters-level and continuing professional development training for the nuclear industries. Since 2003, EPSRC has contributed £6 million towards a research programme which brings together seven universities, various Government bodies, and the private sector, who have also contributed funds. The higher education sector itself has also responded to demand for nuclear specialists and some 11 university-level institutions now offer masters courses in nuclear science or engineering. In addition, EPSRC is inviting proposals for a Centre for Nuclear Engineering under the Engineering Doctorate scheme, with contributions expected from private and public sector partners. This is part of a wider strategy to address potential skills shortages in research.

2.214 Outside of the immediate nuclear industry, the Government is assisting Energy and Utility Skills, the Sector Skills Council for electricity, gas, water and waste management, together with its client employers, to develop a skills strategy for the electricity sector. For new build, engineering construction faces a double challenge of an ageing workforce coupled with a major up-turn in new construction. The Engineering Construction Industry Training Board is working with its employer partners to increase recruitment and training to improve the supply of skills for energy sector projects.

**112**

2.215 The Secretary of State for the then DTI announced in October 2006 that, subject to the agreement of appropriate contractual terms, a National Nuclear Laboratory (NNL) would be established based around the staff at Nexia Solutions and the research facilities owned by the NDA, including the Sellafield Technology Centre.

2.216 The programmes to develop skills within the nuclear industry compare very well with what other parts of the energy sector have done so far. Overall, the nuclear industry is comparatively well-placed to meet the challenges to come.

## *Our conclusion*

**Having reviewed the arguments and evidence put forward, the Government believes that the energy sector, nuclear and otherwise, faces challenges in meeting its need for skilled workers and in the capacity of the manufacturing supply chain to support new construction. However, we believe that the situation is manageable and that building new nuclear power stations does not present a significantly greater challenge than the alternatives. Indeed, a nuclear renaissance, here and around the world, presents opportunities for companies to grow and for individuals to have rewarding careers. We conclude, therefore, that the skills and supply chain situation does not provide a reason to prevent energy companies from investing in new nuclear power stations.**

# Reprocessing of spent fuel

**THE GOVERNMENT'S VIEW**

The Government has concluded that any nuclear power stations that might be built in the UK should proceed on the basis that spent fuel will not be reprocessed and that accordingly waste management plans and financing should proceed on this basis.

**Question 14**
**Do you agree or disagree with the Government's views on reprocessing? Are there any significant considerations that you believe are missing? If so, what are they?**

## *Key arguments and issues presented in responses*

2.217 There was some backing among respondents to the consultation for the Government's position that any new nuclear programme should proceed on the basis that we will not reprocess spent nuclear fuel. However, some people qualified their support and suggested that options should be kept open.

2.218 Some respondents opposed reprocessing as a matter of principle, while some saw it as unnecessary. Others did not think it was sensible to close off options in the light of rising uranium prices and were concerned that a programme of new nuclear power stations would generate a greater volume of waste, in the form of spent fuel, if we were to forego reprocessing.

2.219 Other respondents who supported building new nuclear power stations said that they would prefer to see spent fuel from new reactors being reprocessed. However, they agreed with the Government's approach if it would facilitate investment in new nuclear power stations.

2.220 Some saw reprocessing as a prudent form of recycling a finite resource. They argued that if we can reuse nuclear fuel then we should do so, and that it would be short-sighted to abandon reprocessing.

2.221 A number of people felt that the UK should build on its expertise in reprocessing and that the Government should review how reprocessing and the use of Mixed Oxide (MOX) fuel (fuel made from a mix of plutonium oxide and uranium oxide) could contribute to security of energy supply (by providing another fuel source) and maintain existing skills.

2.222 Some contributors to the consultation observed that potentially reducing the volume of spent fuel by reprocessing could be seen as a benefit by any future host community for a geological disposal facility.

2.223 A small number of people referred to GNEP[207] and to the use of fast-breeder reactors to reduce the volumes of waste and to increase the amount of energy that we could extract by recycling spent fuel. Similarly, some suggested that pebble-bed reactors could lessen the need for reprocessing.

## Government response

2.224 As we set out in the consultation document, reprocessing has advantages and disadvantages. Although reprocessing could make a contribution to security of supply through the creation of raw materials for MOX fuel, as we explain (see paragraph 2.37), we do not believe that uranium resources or the future price of uranium will be limiting factors for new nuclear power stations. The cost of uranium is a relatively small component of the cost of nuclear generation.

2.225 Whether spent fuel from new reactors is regarded as waste or reprocessed, ultimately there will be higher level wastes that will have to be disposed of anyway. Provided we take full account of this when compiling the waste inventory it does not present any new technical challenges in the development of a geological disposal facility. Including spent fuel from new reactors in the waste inventory will impact on the overall size of the geological disposal facility. A significant factor to be considered when emplacing either HLW from reprocessing, or spent fuel in a repository, is that they both generate heat. This means they must be carefully spaced out in order to avoid unwanted heat build up and, in the case of spent fuel, to avoid criticality risks.

2.226 Reprocessing of spent fuel produces a smaller volume of HLW than direct disposal of spent fuel, although it creates seperated uranium and plutonium which have to be effectively managed. To put this in context a geological disposal facility that we would have to build for legacy wastes would need to be about 50% larger to accommodate spent fuel from a programme of 10 new AP1000 reactors[208]. If that fuel were to be reprocessed before disposal, the geological disposal facility for legacy wastes would need to be about 15% bigger to accomodate the additional HLW.

2.227 Longer-term initiatives such as GNEP are at a very early stage and will have no immediate influence on the Government's position on reprocessing. A fast-breeder reactor is unlikely to form part of the nuclear fuel cycle in the short-to-medium term. Likewise, pebble-bed reactors are still at the R&D stage with little sign that they will enter commercial service for many years.

---

207 The Global Nuclear Energy Partnership (GNEP) is an US initiative for an international partnership to enable the expanded use of nuclear energy by using technology to produce non-proliferation benefits, make more effective use of fuel resources through recycling and reduce waste volumes. Further information can be obtained from http://www.gnep.energy.gov.

208 The reference to the Westinghouse AP1000 reactor is purely for illustrative purposes.

## *Our conclusion*

Having reviewed the arguments and evidence put forward, and in the absence of any proposals from industry, the Government has concluded that any new nuclear power stations that might be built in the UK should proceed on the basis that spent fuel will not be reprocessed and that plans for, and financing of, waste management should proceed on this basis.

We are not currently expecting any proposals to reprocess spent fuel from new nuclear power stations. Should such proposals come forward in the future, they would need to be considered on their merits at the time and the Government would expect to consult on them.

# Other considerations

We recognise that making a decision on the potential role of nuclear power is a complex issue, and that there are many issues that need to be considered.

**Question 15**
**Are there any other issues or information that you believe need to be considered before taking a decision on giving energy companies the option of investing in nuclear power stations? And why?**

## *Key arguments and issues presented in responses*

2.228 Respondents to the question about other issues that the Government should consider concentrated on support for increasing energy efficiency, alternative forms of electricity generation, such as distributed generation, and renewables, with some calling for more investment in R&D to bring new technologies to the market. Those who did not think there were any additional issues to be considered tended to support immediate action to make new nuclear power stations a reality, although some qualified this by suggesting that the Government should be proactive in communicating the economic and wider benefits of nuclear power, and the advantages it has over other forms of generation. Others expressed concern that the Government should not expose the taxpayer by subsidising energy companies or rescuing companies in financial difficulties as happened with British Energy. Primarily to address safety and security concerns, some people preferred to see some form of public ownership of nuclear generation. Some also suggested that a new body should oversee a new nuclear programme.

## *Government response*

2.229 Many of the issues raised in response to this question are already addressed at the appropriate points elsewhere in this White Paper and we have therefore only dealt with certain points below. As the Government set out in 2007[209], support for energy efficiency and renewables is central to our strategy to tackle climate change and to increase low-carbon electricity generation in the UK. We have a target that aims to see renewables grow as a proportion of our electricity supplies to 10% by 2010, with an aspiration for this level to double by 2020. The Renewables Obligation (RO) is the main mechanism for ensuring this growth. We are committed to strengthening the RO, increasing the Obligation to up to 20% as and when increasing amounts of renewables are deployed. The RO and the Climate Change Levy exemption is projected to provide annual support of around £1 billion for deployment of renewable electricity in 2010, rising to around £2 billion a year in 2020.

209 Energy White Paper, *Meeting the Energy Challenge*, URN 07/1006, May 2007.

2.230 The Government has set out in its Microgeneration Strategy[210] how it intends to stimulate the growth of distributed and micro-generation by easing planning restrictions for projects and to provide financial support to develop the market. We are also employing measures to encourage the deployment of combined heat and power (CHP), by including an exemption from the Climate Change Levy, improving treatment under Phase II of the EU's ETS, and by providing better planning guidance to ensure wider consideration of the CHP option. The Government has also set a target to require all new homes to have zero $CO_2$ emissions by 2016[211].

2.231 This White Paper sets out how the Government will ensure operators meet their full decommissioning costs and their full share of waste management and disposal costs (see paragraphs 3.46-3.75 in Section 3). It is the Government's policy that the operators of new nuclear power stations must set aside funds over the operating life of the power station to cover the full costs of decommissioning and their full share of waste costs. These financing arrangements must be robust, and designed to deliver sufficient funds even under scenarios such as the insolvency of the power company or the early closure of the power station. The Government is taking powers through the Energy Bill to introduce this financing mechanism.

---

210 Microgeneration Strategy, *Our Energy Challenge: Power from the people*, March 2006.
211 Department for Communities and Local Government, *Building a Greener Future: policy statement*, July 2007.

# Our proposals on nuclear power

## THE GOVERNMENT'S PRELIMINARY VIEW

The Government is not itself proposing to build nuclear power stations.

We have, however, reached the conclusion that private sector energy companies should have the option of investing in new nuclear power stations, subject to the following conditions:

- the developer preparing an Environmental Impact Assessment[212] and securing development consent;
- the developer securing the necessary permissions from the independent regulators to ensure that the nuclear power station could be operated safely, securely and without detriment to public health;
- a decision by the Secretary of State, that the proposed design is Justified (in accordance with the Justification of Practices Involving Ionising Radiation Regulations 2004)[213];
- the proposal being in a site that meets the suitability criteria as identified through a Strategic Siting Assessment. This Assessment would also meet the requirements for a Strategic Environmental Assessment (in accordance with EC Directive 2001/42);
- the establishment, in legislation, of arrangements to protect the taxpayer and ensure that energy companies meet their full decommissioning costs and full share of waste management costs. These would need to be agreed before proposals for new nuclear power stations could proceed. As with the existing nuclear power stations, there is a potential Government liability in accordance with international Conventions to cover third party damages in the unlikely event of a major accident;
- a decision that the management of waste arising from new nuclear power stations would be explored through the Managing Radioactive Waste Safely (MRWS) process.

Within this framework, we think it is likely that energy companies will come forward with proposals for new nuclear power stations, although we cannot predict this with certainty. Their decisions will be affected by their view on the underlying costs of new investments, their expectations of future electricity, fuel and carbon prices, expected closures of existing power stations and the development time for new power stations. We cannot know all of these things today and believe we should reflect this uncertainty by having a diversified approach in our energy policy. This will reduce the risks associated with this uncertainty, for example, by preventing over-reliance on a limited number of technologies.

212 In accordance with the Council Directive of 27 June 1985 on the assessment of the effects of certain public and private projects on the environment (EIA Directive) 85/337/EEC.
213 Justification is a high-level assessment to determine the benefits and detriments associated with a particular class or type of nuclear practice. Before a new class or type of practice can be introduced into the UK, it must be Justified.

**119**

The Government believes that, given the many uncertainties in the energy market over the coming decades, not allowing energy companies the option of investing in new nuclear power stations would increase the risks of not achieving our long-term climate change and energy security goals, and if we were to achieve them, it would be at higher costs.

Having reviewed the evidence, the Government's considered view is that the advantages of giving the private sector the widest choice of investment options, including nuclear power stations, outweigh the disadvantages. Moreover, we believe that through the regulatory protections already in place, and other risk mitigation approaches described in this document, the risks can be effectively managed.

**Question 16**
**In the context of tackling climate change and ensuring energy security, do you agree or disagree that it would be in the public interest to give energy companies the option of investing in new nuclear power stations?**

## Key arguments and issues presented in responses

2.232 Overall, more participants in all strands of the consultation agreed than disagreed that it would be in the public interest to give energy companies the option of investing in new nuclear power stations. The main reason for agreement given by those responding to the consultation document was the proven ability of nuclear power to provide a baseload capacity, which is low-carbon and uses fuel from secure sources. However, many of those at the deliberative and stakeholder events saw nuclear very much as a short-term solution which should be used only until renewables were able to meet the twin challenges.

2.233 At the deliberative events, 44% agreed with the question, with 37% disagreeing and 18% feeling unable to come to a view. Some felt that not enough information had been provided to be able to come to a decision. There was considerable agreement with the Government's view at the stakeholder meetings although, as was seen at the public events, this was sometimes given reluctantly.

2.234 For a number of participants in the stakeholder meetings, nuclear power was a straightforward and obvious solution. It was felt to be counter to diversity of supply principles to remove nuclear power as an option at this stage. Some of those responding to the consultation document wanted nuclear power to represent a larger proportion of the electricity mix than was currently the case. Many wanted to keep the option open but had a number of supplementary views, some of which are outlined below, and more are highlighted in our analysis of responses to Question 17.

2.235 Some people thought that the private sector, with an interest in maximising profits, might cut corners to reduce costs, possibly

compromising safety. Others said that, given the Government's view that it is very important to tackle climate change and ensure security of supply, it is incongruous to leave it to the market to come forward with proposals, or not, as the case may be. In either instance, these respondents were pointing to the need for a stronger role for Government in procuring and operating any new nuclear power stations.

2.236 Other respondents expressed concern that the Government's decision to rely on the market to deliver the most cost-effective low-carbon, high security electricity mix of generation technologies might not lead to any new nuclear power stations.

2.237 Those agreeing with the Government's proposition on new nuclear power stations often also felt that, given the importance of tackling climate change and improving security of supply, it would be too slow to leave market forces to deliver new capacity. Others felt, for similar reasons, that the Government's proposed programme of facilitative actions lacked urgency. Some advocated that the Government should be more interventionist, possibly procuring new nuclear power stations, or offering financial incentives for their construction. In the context of security of supply, the perceived finite availability of uranium was sometimes also cited as a potential problem.

2.238 Some respondents expressed concerns about safety, security, health and the environment. Some who agreed with the Government's preliminary view on nuclear power said that if we do build new nuclear power stations, we must address these issues properly as an integral part of any new programme. Those who disagreed with the Government's preliminary view often cited these issues as specific reasons why we should not allow the construction of new nuclear power stations.

2.239 Some people felt that allowing new nuclear power stations to go ahead would divert resources from alternative low-carbon electricity technologies, such as renewables or carbon capture and storage. Some viewed this as a reason in its own right not to allow the option of investment in new nuclear power stations. Others felt that if we gave energy companies the option to build new nuclear power stations, then the Government should make additional efforts to preserve or encourage alternatives, alongside any nuclear new build.

2.240 Some respondents also mentioned waste as a serious issue that required the Government's attention. This view came from both those who agreed with the Government's preliminary view on allowing the option of new nuclear build and those who disagreed.

2.241 Smaller numbers of respondents expressed a wide spectrum of other views. These are explained in the accompanying analysis document[214], and the key views are also discussed in relation to Question 17.

---

214 The Future of Nuclear Power, *Analysis of consultation responses*, URN 08/534, January 2008.

2.242 The Scottish Executive has made clear its opposition to the UK Government's proposals on new nuclear power. They note that any application for development consent for new nuclear power stations in Scotland would require the consent of Scottish Ministers under section 36 of the Electricity Act 1989, and say that while any proposal from the industry to build a new nuclear power station would need to be considered on its individual merits it is unlikely that such proposals from industry would find favour with the Scottish Executive.

2.243 The Welsh Assembly recognises the role that existing nuclear power stations in Wales play in the generation of electricity and their role in the local economy. The Welsh Assembly have said that given current or planned energy projects scheduled to come on stream in the next 15 years, it considers that pursuing new nuclear power stations in Wales is unnecessary.

## Government response

2.244 It remains a central plank of the Government's energy policy that competitive energy markets, with independent regulation, are the most cost-effective and efficient way of generating, distributing and supplying energy, to meet the twin challenges of tackling climate change and ensuring energy security.

2.245 In this context, we have concluded that it would be in the public interest to allow energy companies, not the Government, the option of investing in new nuclear power stations. In reaching this conclusion we have carefully considered the evidence and arguments set out in the consultation document and have considered the responses to the consultation and any other relevant evidence which has emerged. In particular, we have considered a range of issues including:

- nuclear power and carbon emissions
- security of supply impacts of nuclear power
- the economics of nuclear power
- the value of having low-carbon electricity generation options: nuclear power and the alternatives
- the safety and security of nuclear power
- transport of nuclear materials
- waste and decommissioning
- nuclear power and the environment
- the supply of nuclear fuel
- supply chain and skills implications
- reprocessing of spent fuel

2.246 These issues are discussed elsewhere in Section 2 of this White Paper. Having considered the issues in the round, we continue to believe that we face two long-term challenges, namely tackling climate change by reducing $CO_2$ emissions both in the UK and abroad, and ensuring the security of our energy supplies.

2.247 There is also considerable uncertainty about the future energy mix, in particular, the pace of climate change and the pressures this will create,

and geopolitical developments. There are also uncertainties relating to future fossil fuel and carbon prices; the speed at which we can achieve greater energy efficiency and therefore likely levels of energy demand here and globally; the speed, direction and future economics of development in the renewable sector; and the technical feasibility and costs associated with applying carbon capture and storage technologies to electricity generation on a commercial scale.

2.248 In view of the need to achieve our twin energy challenges and given the uncertainties about the future energy mix, we believe that preventing energy companies from investing in new nuclear power stations would increase the risk of not achieving our long-term climate change and energy security goals, or achieving them at higher cost.

2.249 However, we recognise that there are significant concerns about a number of issues associated with nuclear power. For example, the public are concerned about risks in relation to safety, security, proliferation, transport and the environment. Whilst these are understandable concerns we believe that the risks associated with nuclear power are small and that the existing regulatory regime is such that those risks can be effectively managed.

2.250 The public is also concerned about the management of radioactive waste. We recognise the importance of having a mechanism for the long-term management of radioactive waste. We are satisfied that it would be technically possible to dispose of new nuclear waste in a geological disposal facility and that this waste could be stored safely and securely until such time as the geological disposal facility is ready. We are exploring that mechanism through the MRWS process and believe it will provide a feasible mechanism for identifying a suitable site for construction of the geological disposal facility. We set out further views on this in this Section where we deal with waste and decommissioning.

2.251 We recognise that there are also other concerns including concerns about the supply of uranium, skills and about the environmental impact of nuclear power. Whilst we accept that these are important issues, we think these issues can be managed and as such, we do not think they provide a reason for not allowing energy companies to invest in new nuclear power stations.

2.252 Having considered the issues above and the other arguments and evidence raised in the consultation and in the responses to it, we have concluded that it would be in the public interest to allow energy companies the option of investing in new nuclear power stations. The next steps the Government will take to facilitate investment in new nuclear power stations are outlined in Section 3 of this White Paper. An analysis of views on the restrictions which might be applied to the construction of any new nuclear power stations is included in the analysis of responses to Question 17 of the consultation.

2.253 Nuclear energy is largely a reserved matter, however, as set out in the 1999 Concordat between DTI (as it then was) and the Scottish

**123**

Executive[215], "certain functions relating to energy matters have been 'executively devolved' to Scottish Ministers, enabling them to take certain decisions on energy matters within the framework of UK Energy Policy". The power to consent to the construction of power stations greater than 50MW capacity has been executively devolved to Scottish Ministers and is also devolved in Northern Ireland. In developing the proposals set out in this White Paper we will take account of any areas in which the Devolved Administrations have competence.

## Our conclusion

**In the context of tackling climate change and ensuring energy security, the Government has concluded that it would be in the public interest to give energy companies the option of investing in new nuclear power stations.**

215 Concordat between the DTI (now BERR) and the Scottish Executive dated 25th November 1999.

# Other conditions

Immediately after asking the "in principle" question (Question 16) we also consulted on the proposition that conditions might be attached to any new build programme.

**Question 17**
**Are there other conditions you believe should be put in place before giving energy companies the option of investing in new nuclear power stations? (For example, restricting build to the vicinity of existing sites, or restricting build to approximately replacing the existing capacity.)**

## *Key arguments and issues presented in responses*

2.254 In answering this question, and in answers to other questions, people attached a number of conditions to both agreement and disagreement with the Government's views. However, a substantial number of people answering this question said that no further specific restrictions, beyond the normal regulatory processes already in place, should be applied to potential investors in new nuclear new power stations. Conversely, others said that restrictions were irrelevant because companies should not be allowed the option of building new nuclear power stations in the first place.

2.255 When conditions were attached, the key conditions which were consistently mentioned across all strands of the consultation were the need to have:
- reassurance that investment in renewables and demand side (energy efficiency) initiatives would be protected and, in many cases, increased. Some of those at the stakeholder meetings felt that the Renewables Obligation would help to guarantee future investment in renewable solutions; others at the deliberative events were also aware of the Obligation and wanted to see more investment in the shorter-to-medium-term
- a strong regulatory framework and an ongoing programme of independent scrutiny and inspection
- reassurance that the Government will act to ensure that the private sector is accountable, most importantly in adhering to strict safety standards and bearing full costs
- guarantees that robust plans for the long-term management of waste are in place before the construction of new plants
- protection for the taxpayer against subsidising the private sector, most crucially behind waste management and decommissioning
- ways of minimising the environmental impact by restricting new build to existing sites; this was also seen by some of those responding to the consultation document as having the benefit of speeding up planning approval and construction timings, as well as utilising existing infrastructures

2.256 In addition to these consistently-mentioned concerns, there were a number of other points made by smaller numbers of participants in the consultation. These are outlined in the following paragraphs.

2.257 There was some discussion at the deliberative events about how many new nuclear power stations should be built. Some felt that this should be restricted to the number of power stations we currently have, others that an unspecified cap should be imposed. Some people at both the deliberative and stakeholder events thought that a maximum percentage of electricity generated by nuclear should be defined, with others feeling that as there is a threat of the UK being unable to meet its future energy needs, no restrictions should be imposed.

2.258 There was some concern amongst those responding to the consultation document that existing sites (which are located on the coast) may not be suitable in the future due to increased risks of flooding as sea levels rise. This led some to feel that developers should be able to identify the most appropriate locations, rather than being constrained to existing sites. A number of people again raised the need for local communities to be involved in the planning process and felt that there should be compensation for any communities that do host new nuclear power stations.

2.259 Some of those responding to the consultation document raised the question of ownership as a condition, specifically that investment should only be allowed by British companies or that public ownership would be preferred.

2.260 Some additional conditions raised specifically by those attending the deliberative events included the need to set up a decommissioning fund to protect the taxpayer. A general request was made that all other possible steps to tackle climate change are taken in parallel at home and abroad, and that other countries should be encouraged to act similarly. Finally, the perceived inherent risks in new nuclear power stations, some felt, drives a need for more open communication and education, which would also help to address the concerns of some stakeholders about public misconceptions.

2.261 Some of those attending the stakeholder meetings suggested more detailed conditions relating to the role of the Government. Some felt that the Government appears to be shying away from its responsibility to ensure a diverse energy mix, which was seen as unethical to leave to market outcomes. A few expressed the need for ongoing state involvement in the energy industry, citing countries like France as an example, whilst others wanted more longer-term signals to be sent to the market. Some also felt that the Government has a key role to play in any future carbon pricing mechanisms.

2.262 A number of people did observe that, in their view, any new build programme should be seen as the last, on the assumption that in due course, energy efficiency, progress with renewables and possibly also nuclear fusion would render an open-ended programme redundant.

## *Government response*

2.263 The specific conditions raised by those responding to the consultation were diverse. Many of the issues raised are already addressed at the appropriate points elsewhere in this White Paper and we have therefore only dealt with certain points below.

2.264 There was no clear consensus about the need either to restrict new build to the vicinity of existing sites – though many respondents thought that this would be likely to happen naturally anyway – or to restrict new build to approximately replacing existing capacity. On the latter point, the Government has therefore decided that no specific cap on future new nuclear capacity should be applied.

2.265 We expect that applications for building new power stations will focus on areas in the vicinity of existing nuclear facilities. Industry has indicated that these are the most viable sites. The suitability of sites will be assessed through the forthcoming SSA process. In addition to the SSA, any developer wishing to construct a new nuclear power station would also need to obtain relevant environmental, health and safety authorisations as well as development consent. We will consult on the criteria for assessing suitable sites and then on a draft list of sites. The Government will continue to monitor whether an appropriate market in suitable sites is developing.

2.266 We do not think it is appropriate to restrict new build to approximately replacing existing capacity because the fundamental principle of our energy policy is that competitive energy markets, with independent regulation, are the most cost-effective and efficient way of generating, distributing and supplying energy. In those markets, investment decisions are best made by the private sector and independent regulation is essential to ensure that the markets function effectively.

2.267 We have also considered whether there is any need to impose any other restrictions on new build and have considered the comments made in response to the consultation. Many of the comments made have been addressed elsewhere in this White Paper and we do not address them all specifically here.

2.268 We have, however, considered whether it is necessary to take additional steps to promote investment in renewables, alongside nuclear. On this matter, we have concluded that our plans to extend the RO level up to 20%, subject to deployment, and to target additional support to help bring emerging technologies such as offshore wind and marine to market quicker, will adequately address this concern.

2.269 We have also considered the concerns about ensuring that energy companies adequately provide for waste and decommissioning costs. This is why, in addition to the measures we will be taking in the forthcoming Energy Bill, we have decided to create a Nuclear Liabilities Financing Assurance Board (NLFAB) as explained in Annex C and in Box 4.

## Our conclusion

We are taking steps to facilitate nuclear new build as outlined in this White Paper. In addition we are setting up the Nuclear Liabilities Financing Assurance Board (NLFAB), putting in place measures to ensure that the effectiveness of the Nuclear Installations Inspectorate is further enhanced, and reforming the planning system.

We think the Strategic Siting Assessment (SSA) and Strategic Environmental Assessment (SEA) processes will enable suitable sites to come forward. The Government will continue to monitor whether an appropriate market in suitable sites is developing. The Government expects that applications to build new nuclear power stations will focus on areas in the vicinity of existing nuclear facilities. However, we do not consider it is necessary to put in place additional restrictions or conditions before giving energy companies the option of investing in new nuclear power stations.

# Our proposals for facilitative action

## THE GOVERNMENT'S PRELIMINARY VIEW

If we conclude that energy companies should be allowed to invest in new nuclear power stations, the Government would carry out a package of facilitative actions designed to reduce the regulatory and planning risk associated with investing in nuclear power stations.

The package of measures is designed to reduce the uncertainties in the pre-construction period for new nuclear power stations through improvements to the regulatory and planning processes. The measures will also set out arrangements for the funding of decommissioning and waste management and disposal. The proposed package of measures covers:

- taking steps to improve the process for granting planning consent for electricity developments by ensuring it gives full weight to national, strategic and regulatory issues that have already been the subject of discussion and consultation. This could take the form of a National Policy Statement, consistent with the reforms proposed in the 2007 Planning White Paper[216]. We would:
  - develop criteria for suitable sites for new nuclear power stations through a Strategic Siting Assessment, subject to relevant European and domestic legislative requirements; and
  - continue our consideration of the high-level environmental impacts through a formal Strategic Environmental Assessment in accordance with the SEA Directive[217]. Applicants for specific proposals would still need to carry out a full Environmental Impact Assessment;
- running a process of "Justification" (in accordance with the Justification of Practices Involving Ionising Radiation Regulations 2004);
- the nuclear regulators pursuing a process of Generic Design Assessment[218] of industry preferred designs of nuclear power stations to complement the existing licensing processes. This would consist of an assessment of the safety and security of power station designs and their radiological discharges to the environment; and
- developing arrangements that would protect the taxpayer by ensuring that private sector operators of nuclear power stations securely accumulate the funds needed to meet the full costs of decommissioning and full share of waste management costs. This would need to be agreed before proposals for new nuclear power stations could proceed.
- the power to consent to the construction of power stations greater than 50MW capacity has been executively devolved to Scottish Ministers and is also devolved in Northern Ireland. In developing the proposals above we will need to take account of any areas in which the Devolved Administrations have competence.

---

216 Planning White Paper, *Planning for a Sustainable Future*, May 2007.
217 Directive 2001/42/EC of 27 June 2001 on the assessment of the effects of certain plans and programmes on the environment (O.J. L197, 21.7.2001, p 30).
218 This is sometimes referred to generically as "pre-licensing".

**Question 18**
**Do you think these are the right facilitative actions to reduce the regulatory and planning risks associated with such investments? Are there any other measures that you think the Government should consider?**

## Key arguments and issues presented in responses

2.270 By far the greater number of respondents answering this question supported the facilitative actions we proposed in our consultation. Most urged the Government to proceed as quickly as possible. Most of those of who responded to this question focused on Justification and SSA rather than GDA and waste which are dealt with in Section 3 of this White Paper. A number wanted the Government to be proactive in making the case for nuclear by raising awareness of its benefits. In one or two cases respondents did not think the Government had gone far enough to attract investment in nuclear power and called for further streamlining of the regulatory steps. They did not, however, specify what they felt was necessary, other than stopping further rounds of consultation.

2.271 There was also some concern over what role and influence Devolved Administrations might have in determining whether and where we can build new nuclear power stations. A significant proportion of those who disagreed with the Government did so from a position of general opposition to building any new nuclear power stations.

2.272 The remainder of the respondents objected to the proposed measures to improve the energy planning system for nuclear power stations because they felt they would deny adequate local scrutiny. Some saw the proposals on planning as tilting the playing field in favour of applications for new nuclear power stations and called for a tightening rather than what they saw as a relaxation in the current system of planning consents. Similar concerns about local democracy and public participation in relation to nationally significant infrastructure projects had been raised by people responding to the Government's consultation on the reform of the planning system. Others wanted to see action to guarantee or underpin the carbon price and for the National Grid infrastructure to be reviewed to meet the requirements of modern nuclear power stations.

2.273 Alongside our nuclear consultation, we also consulted on proposals for Justification and a combined SSA and SEA. Because they form part of the facilitative actions we had proposed in forming our preliminary view, we have included an assessment of responses in this White Paper (see Annex B).

2.274 On Justification, a number of respondents felt that the Government's proposals were sufficiently transparent and robust and supported the information requirements we had proposed in the technical consultation document[219]. Some felt that the threat of a terrorist strike should be factored into any safety scenarios considered, along with issues such as waste disposal routes and resource levels. However, some felt that the process was unnecessarily elaborate and emphasised that Justification was meant to be a high-level process with more extensive consideration of the detailed aspects of new build being covered as part of the licensing and planning processes. They argued that in other European Union Member States, Justification was considered as part of other regulatory processes and did not require a separate application. Some suggested that the timeframe was either too long, or should be strictly adhered to.

2.275 There was significant support for the Government's proposal for multiple technologies to be considered under a single process of Justification. Some respondents, however, felt that each technology should be subject to a separate decision and that the process would need to be carefully managed to reduce the prospect of delays.

2.276 In respect of our proposals for a SSA and SEA, most respondents to the consultation supported the proposed three-stage process. They agreed that it is a logical and robust process for assessing sites. There was also agreement that we should incorporate the SEA into the SSA and that this is the correct approach to assessing environmental impacts at a strategic level. In addition, the following views were expressed:
- The SSA should prioritise existing sites (for example, because of local support and established connections to the National Grid; and because Environmental Impact Assessments (EIAs) already exist for many sites such as Hinkley C, Wylfa B and Sizewell C)
- Stages 1 and 2 of SSA could proceed concurrently
- There is a need to integrate Justification and SSA processes adequately
- There is a need to use nuclear experts to ensure that the SSA is technically adequate
- The threat of climate change should rule out existing sites by the coast
- SSA should address climate change and the threat of earthquakes
- Concern that the existing grid rules limits, as embodied in the GB Security and Quality of Supply Standard (GBSQSS[220]), is a barrier to the development of nuclear power plants with generating units exceeding 1320MW, and that Government should review the GBSQSS as part of the facilitative actions[221].

219 The Future of Nuclear Power, *The Role of Nuclear Power in a Low Carbon UK Economy, Consultations on the proposed processes for Justification and Strategic Siting Assessment,* URN 07/972, May 2007.
220 GB Security and Quality of Supply Standard, Issue 1, 22 September 2004.
221 The GBSQSS sets out a set of criteria and methodologies that the GB Transmission Licensees are required to use in the planning and operation of the GB Transmission System.

## Government response

2.277 We are encouraged by the support for the facilitative steps we have proposed. We believe these steps constitute a coherent and comprehensive package to remove uncertainties and inefficiencies.

2.278 We have looked again at the Justification process to assess whether the process could be more efficient, whilst ensuring it remains fit for purpose. We are also introducing measures to ensure that the process can be managed effectively so that the timeframe does not slip. (We set out these measures in Section 3 of this White Paper.)

2.279 In terms of the approach taken by other Member States, we believe that the Justification process we have developed for the UK meets the requirements of the regulations, is transparent, efficient and robust and is based on existing regulations[222]. When it comes to fast-tracking technologies that have already been Justified by other EU Member States, the Justification Authority must consider new classes or types of practice on their merits. We are bound by both European and UK Law which state that any new class or type of practice needs to be Justified for use by the relevant Justifying Authority.

2.280 To provide clarity on the definition and treatment of health detriments and practices, we will provide more information in the detailed guidance being prepared on the process of Justification for new nuclear power stations. These are designed to provide general guidance to those considering making an application and will be used in conjunction with the general Justification Guidelines[223] produced by Defra. We will issue these alongside a call for Justification applications.

2.281 We acknowledge the concerns raised regarding the grid rules limits, as set out in the GBSQSS. The Government will discuss with Ofgem, the relevant regulatory authority, on how to take this concern forward, but it is too early to say how this issue should be resolved before further analysis of the potential impact of new reactors on the transmission system takes place. The issue of allocation of costs arising from any changes is a matter for Ofgem to review with the industry. However, the industry and National Grid should consider together the impact of larger generating units on the transmission system and appropriate approaches to dealing with that impact. This consideration may include a review of the relevant rules contained in GBSQSS.

2.282 We recognise the strength of feeling that exists over the Government's proposed reform of the planning system for major infrastructure projects, and which was reflected in the responses to the Planning White Paper. The Government took these into account in developing the Planning Bill, which seeks to strengthen and improve the proposals for reform in the Planning White Paper to ensure accountability, public participation and promote a sustainable approach to development. As we explain further (see Section 3), a National Policy Statement (NPS) covering nuclear power would cover the development consents for

---

222 Justification of Practices Involving Ionising Radiation Regulation 2004 (S.I. 2004/1769).
223 www.defra.gov.uk/environment/radioactivity/Government/legislation/justification.htm

new nuclear power stations. The NPS would set out the Government's policy on the national strategic issues which need to be taken into account when granting consent to the construction of any new nuclear power stations. The NPS would to a large extent build on the proposed SSA. As mentioned above, the SSA will involve consultation with local communities.

2.283 We set out details of how the Government will take forward these facilitative steps in Section 3.

# What the Government will do

3.1 The Government has reached the conclusion that new nuclear power stations can help the UK to meet its objectives on climate change and energy security. We conclude, therefore, that it would be in the public interest to allow energy companies the option to invest in new nuclear power stations. The Government will take a number of facilitative actions to reduce regulatory and planning risks associated with investing in new nuclear power stations and to ensure that owners and operators of new nuclear power stations set aside funds over the operating life of the power station to cover the full costs of decommissioning and their full share of long-term waste management and disposal costs. These facilitative steps, listed at paragraph 3.5, will reduce uncertainties in the pre-construction period through improvements to the regulatory and planning processes.

3.2 This Section sets out information on:
- what facilitative action the Government proposes to take
- information on the specific proposals

3.3 Nuclear power stations have long lead times and require major capital investment. To proceed in a competitive market for energy, investors have to be confident that the regulatory requirements are clear and that decisions will be timely. Having concluded that energy companies should be allowed the option to invest in new nuclear power stations, the Government believes it is necessary to undertake facilitative action before it is likely that energy companies would bring forward proposals.

## What facilitative action is the Government proposing?

3.4 Government's intention is that the inquiry phase of any application for a new nuclear power station should examine the proposal in the context of the national strategic or regulatory material considerations, which will already have been established through our facilitative action. It should examine the local benefits of the development and how local impacts of the construction and operation of the plant can be minimised. The purpose of our facilitative action is therefore to handle these national strategic and regulatory material considerations and enable the consideration of the proposal to progress effectively and efficiently.

3.5 The facilitative action we propose to take is designed to reduce the regulatory uncertainty and risk associated with investing in new nuclear power stations by:
- Improving the planning system for major electricity generating stations in England and Wales, including nuclear power stations, by ensuring it sets a framework for development consents that gives

**134**

full weight to policy and regulatory issues that have already been subject to debate and consultation at a national level, and does not reopen these issues in relation to individual applications

- Running a SSA process to develop criteria for determining the suitability of sites for new nuclear power stations. Subject to some European legislative requirements, this would enable the planning process to focus on the proposals rather than debate whether there are other more suitable sites for development

- In conjunction with the SSA, taking further our consideration of the high-level environmental impacts in accordance with the Strategic Environmental Assessment (SEA) Directive[224]. This would limit the need to consider such high-level environmental impacts of nuclear power stations during the planning process

- Running a process of Justification (in accordance with the Justification of Practices Involving Ionising Radiation Regulations 2004) to test whether the economic, social or other benefits of specific new nuclear power technologies outweigh any health detriments

- Assisting the nuclear regulators to pursue a process of Generic Design Assessment[225] of industry preferred designs of nuclear power stations, to complement the existing site-specific licensing process. This would involve assessing the safety, security and environmental impact of nuclear power reactor designs, including waste arisings and radioactive discharges to the environment. This would limit the need to discuss these issues in depth during the site-specific licensing process

- Working with the regulators to review the regulatory regime to explore ways of enhancing its effectiveness in dealing with the challenges of new nuclear power stations

- Pushing for a strengthening of the Emissions Trading Scheme so that investors have confidence in a continuing carbon price signal when making a decision.

3.6    In addition, through the Energy Bill, we are introducing legislative arrangements to ensure that operators of new nuclear power stations have secure financing arrangements in place to meet the full costs of decommissioning and their full share of waste management costs.

## Potential path to new nuclear power stations

3.7    A potential path, including a timeframe for the facilitative actions we envisage, is set out in Chart 3.

3.8    Our indicative timetable shows, and as we explain later in this section, the first phase of the Generic Design Assessment (which is on the critical path) began, on a contingent basis, in July 2007. This and subsequent phases of GDA are expected to take 3-3½ years to complete. Although this may be subject to some change, depending on the new planning proposals before Parliament, operators are likely

224 Directive 2001/42/EC of 27 June 2001 on the assessment of the effects of certain plans and programmes on the environment (O.J. L197, 21.7.2001, p30).
225 This is sometimes referred to generically as "pre-licensing".

## Chart 3: Indicative pathway to possible new nuclear power stations

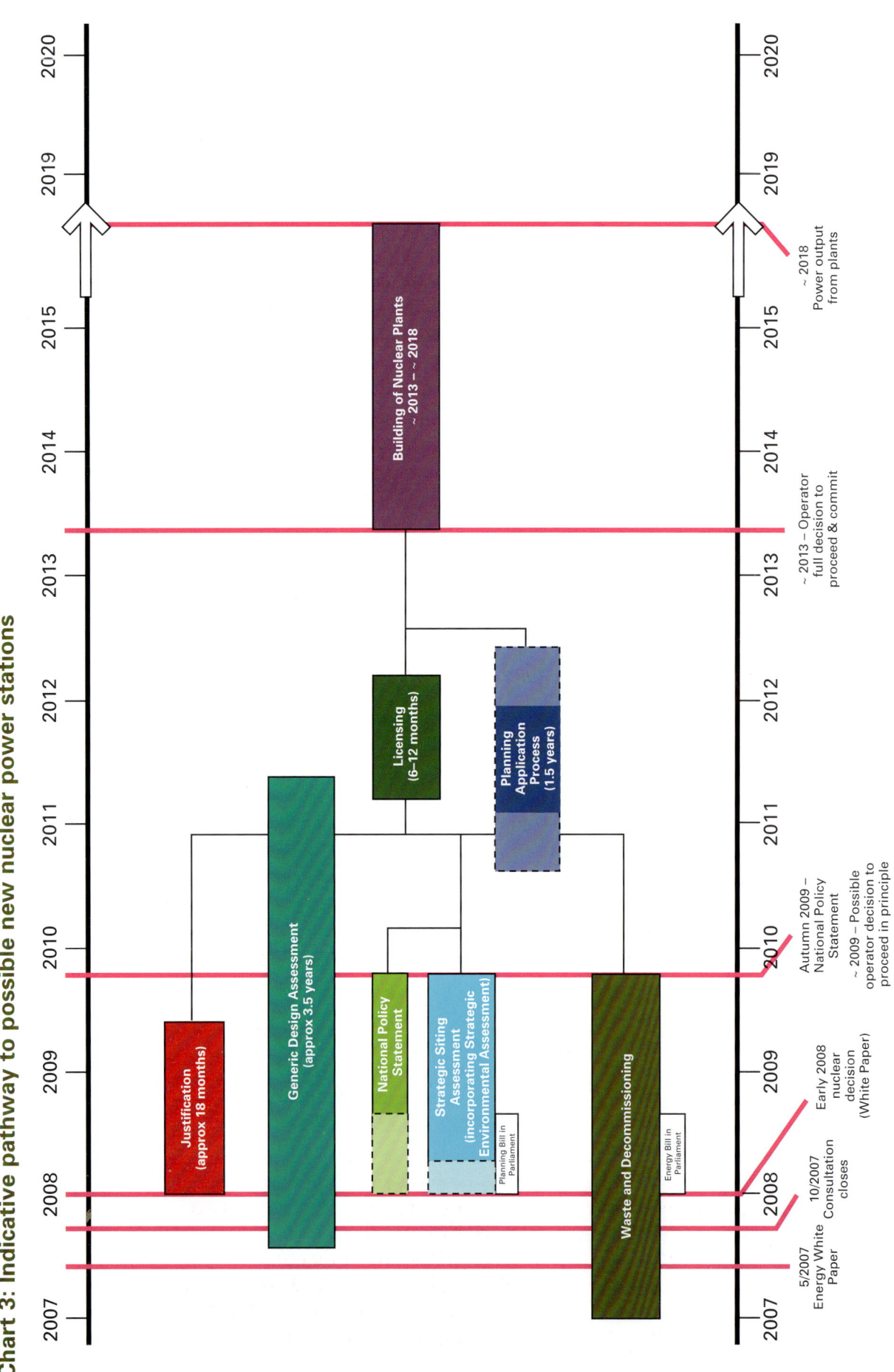

**Justification**
(approx 18 months)

**Generic Design Assessment**
(approx 3.5 years)

**National Policy Statement**

**Strategic Siting Assessment**
(incorporating Strategic Environmental Assessment)

Planning Bill in Parliament

**Waste and Decommissioning**

Energy Bill in Parliament

**Licensing**
(6–12 months)

**Planning Application Process**
(1.5 years)

**Building of Nuclear Plants**
~ 2013 ~ 2018

~ 2018
Power output from plants

~ 2013 – Operator full decision to proceed & commit

Autumn 2009 – National Policy Statement

~ 2009 – Possible operator decision to proceed in principle

Early 2008 nuclear decision (White Paper)

10/2007 Consultation closes

5/2007 Energy White Paper

to be able to make site specific applications in 2010 with construction beginning from 2013/2014 onwards.

3.9    The following indicative table shows the immediate actions the Government will be taking following the publication of this White Paper:

| | |
|---|---|
| Energy Bill containing clauses on nuclear waste and decommissioning financing provisions | January 2008 |
| Consultation on waste and decommissioning financing provisions guidance | February/March 2008 |
| Consultation on SEA scope | February/March 2008 |
| Call for applications and publication of Justification Guidance | March 2008 |
| Consultation on draft SSA criteria | March/April 2008 |
| Announce start of process for selecting reactor designs for next stage of GDA | Spring 2008 |
| Government full response to MRWS consultation | Spring 2008 |

## The Planning Bill reforms and new nuclear power stations

3.10   The Government recognises the impact that an effective planning system could have on successfully and fairly achieving our energy goals. In 2007[226], the Government published proposals for a fundamental reform of the planning system for nationally significant major infrastructure projects and in November 2007 introduced a Planning Bill to bring about this reform, taking account of responses received in the course of the consultation on the original proposals. The proposals below, in particular in relation to the National Policy Statement (NPS) and the Infrastructure Planning Commission (IPC), are based on the assumption that the proposed planning reforms proceed. The precise way forward will need to take account of any amendments to the Planning Bill. In the event that there are delays to the implementation of the reforms to the planning system, we will consider other options to make clear the national case for new nuclear power stations.

3.11   In the past, the planning process for nuclear power stations has been inefficient, costly and lengthy, and, in some cases may not have provided sufficient opportunity for consideration of local issues because they spent much of their time dealing with broader national issues. For example, the nuclear power station Sizewell B took six years to secure planning consent, costing £30 million, and only 30 of the 340 inquiry days were devoted to local issues. The planning reforms to be introduced through the Planning Bill increase transparency and

226 Planning White Paper, *Planning for a Sustainable Future*, May 2007.

participation and will deliver a number of important improvements to the planning process for the development of nationally significant infrastructure, including energy projects, such as new nuclear power stations.

3.12 The Planning Bill will establish a new single consent regime for nationally significant infrastructure under which the Government would produce NPSs that will establish the national case for infrastructure development and set the policy framework for IPC decisions. These NPSs will be subject to public consultation and the Planning Bill sets out the procedures for producing them. Decisions on applications will be made by an independent IPC which will manage inquiries and take decisions on applications for nationally significant infrastructure projects, including proposed new nuclear power stations.

3.13 We are proposing that there will be a National Policy Statement which would address nuclear power. This Nuclear NPS would reiterate the Government's policy on nuclear power, building on this White Paper. A key component of the NPS which covers nuclear power will be siting criteria which the Government considers should be used to assess the suitability of potential sites for nuclear power stations, and an indication of certain locations that met these criteria following a SSA.

3.14 If the NPS which covers nuclear power includes an indication of which locations may be suitable for new nuclear build, we envisage that there would be a process of engagement and consultation with those local communities on which the NPS had a direct bearing, before this was finally adopted. We would expect this to be conducted as part of the SSA process which is described further below. This would be consistent with the requirement proposed in the Planning Bill that where any NPS identifies specific locations, the Secretary of State must ensure appropriate steps are taken to publicise the proposal. The planning reforms would also create an active pre-application phase, during which potential developers will need to consult publicly and locally on their proposals and engage with local authorities, statutory bodies and other key parties before submitting their application to the new IPC.

3.15 The IPC's responsibilities will cover England and Wales for power generation, and NPSs are not expected to have statutory force beyond England and Wales. However, the policy set out in the NPSs relating to energy will be developed within the framework of UK energy policy and with reference to the whole of Great Britain or UK as appropriate not just England and Wales. Decisions on development consents for nuclear power stations will continue to be taken in Scotland by Scottish Ministers who act within the framework of UK energy policy.

3.16 The proposed NPS which would include nuclear power would therefore provide the framework for the IPC to take decisions on applications to build new nuclear power stations in England or Wales. This NPS would draw on the statements in this White Paper which take account of economic, environmental and social considerations in relation to nuclear

power. It would also include the outcomes of the proposed SSA in terms of the SSA criteria and the list of sites.

3.17 Welsh Ministers will be statutory consultees for energy NPSs. We will consider further what role the other Devolved Administrations should have in the development of this NPS, and the possible extent to which the policy set out in the NPS may have application outside of England and Wales, consistent with the terms of the existing devolution settlement. This would include, for example, considering the potential relevance of the siting criteria in Scotland and Northern Ireland.

## Strategic Siting Assessment and the Strategic Environmental Assessment

3.18 A number of issues relating to the siting of nuclear power stations are national in nature rather than site specific. For example, there are technical safety issues relating to siting that are reflected in the HSE's Safety Assessment Principles, which are set out at a national level. There are also a range of strategic factors that go beyond technical safety considerations, including over-arching environmental considerations and relevant infrastructure requirements (e.g. road, rail and other transport networks and grid connections to nuclear power stations).

3.19 In order to ensure that these national issues are considered at the appropriate level, the Government will carry out a SSA to identify criteria that will be used to assess the suitability, at a strategic level, of appropriate sites nominated for new nuclear power stations, and to assess the high-level environmental impacts of building on those sites. These criteria will be published in draft and will be subject to public consultation. Following consultation, the Government will assess appropriate nominated sites against these criteria. A further public consultation would then be held inviting views on those nominated sites judged by the Government to meet the criteria. It is envisaged that the SSA will run from early 2008 to mid-late 2009.

3.20 We would expect the criteria and sites identified through the SSA to form a key part of the Government's National Policy Statement which covers nuclear power, which in turn will set the policy framework for planning decisions on nuclear power stations by the new independent Infrastructure Planning Commission. Sites that meet the SSA criteria will still need to comply separately with the site-specific requirements of the nuclear safety regulators and with all relevant environmental regulations.

3.21 As part of the SSA, we intend to conduct a SEA under the Strategic Environmental Assessment Directive[227]. At the start of the SSA process, we will consult appropriate bodies on the scope of the proposed SSA in accordance with the legislation which implements

227 Directive 2001/42/EC of 27 June 2001 on the assessment of the effects of certain plans and programmes on the environment (O.J. L197, 21.7.2001, p.30) implemented by the Environmental Assessment of Plans and Programmes regulations 2004 (S.I. 2004/1633).

**139**

the SEA Directive. The SSA would not, however, replace mandatory assessments for individual projects, including Environmental Impact Assessment (EIA) under European law or those required by the Environment Agencies, the Health and Safety Executive and other regulatory authorities. These would have to be undertaken as part of any application to build a new nuclear power station. So although a site may satisfy the broad SSA criteria, detailed examination of the site may raise issues which would rule out its suitability.

## Stages, key activities and indicative timescales in the SSA and SEA process

### Pre-Stage 1 (Feb/March-April 08)

#### Consultation on SEA scoping report
- Publication of draft scoping report as a basis for consultation on the scope of the Strategic Environmental Assessment (SEA) with the designated SEA consultation bodies and authorities
- Consideration of comments from consultation bodies and authorities
- Finalisation of scope of SEA.

### Stage 1 (March/April-September 08)

#### Consultation on draft criteria and publication of first Environmental Report
- Public consultation focusing on the development of the following two categories of criteria:
  (i)  exclusionary criteria, which will help to rule out those areas unsuitable for new nuclear power stations and highlight those within which there might be suitable sites. Such criteria may include "population density" criteria, for example, and
  (ii) discretionary or detailed criteria for assessing the suitability of possible sites. The application of such criteria will highlight sites that it may be more appropriate to avoid where there are better alternatives, or where significant extra expenditure would be needed to address specific inherent defects in a site
- Publication of a first outline Environmental Report, in accordance with the SEA Directive (with an updated Environmental Report possibly in Stage 2 and a final version to be published at Stage 3)
- Consideration of all responses to the consultation to inform a Government statement on the criteria which will be included in a NPS on nuclear power, as appropriate.

**140**

*Stage 2 (October 08-Feb 09)*

## Invite nominations for potential sites to be considered in the SSA

- Publication of a Government statement on:
    (i) the exclusionary and discretionary criteria to be used for the SSA
    (ii) an indication of broad geographical areas which have been ruled out, in whole or in part, and those areas within which there might be sites suitable for new nuclear power stations, explaining how environmental considerations have been taken into account
    (iii) the invitation to nominate sites within areas not ruled out by the exclusionary criteria, which may be suitable for new nuclear power stations
- At this stage, we would also expect to publish an updated Environmental Report as necessary
- Guidance on the nominations process will be provided in due course
- Sites that are nominated would be assessed against the SSA criteria.

*Stage 3 (Feb/March 09 – September/October 09)*

## Consultation on draft list of nominated sites

- We expect to propose and consult on a draft list of nominated sites
- At the same time we would expect to produce the final Environmental Report
- At the end of the process, we would expect to produce a statement setting out:
    (i) the Government's statement on the criteria
    (ii) the final, non-exhaustive, list of sites which meet the criteria
- At this stage we would expect to produce the Environmental Statement for the purposes of the SEA
- Assuming the planning reforms proceed, we would expect the outputs of the above to be a key element of the NPS

3.22 We would consult on the Environmental Report as required by the SEA Directive and implementing legislation.

3.23 Whilst nuclear energy is largely a reserved matter, the power to consent to the construction of power stations greater than 50MW capacity has been executively devolved to Scottish Ministers and is also devolved in Northern Ireland. In developing the proposals set out in this White Paper we will take account of any areas in which the Devolved Administrations have competence.

3.24 We will therefore discuss with the Devolved Administrations whether, and how, the SSA should extend to Scotland and Northern Ireland, and this will include considering whether to develop SSA criteria that could have UK-wide coverage.

**141**

3.25  The Welsh Assembly Government does have some devolved planning functions, but these do not include the power to consent to the construction of power stations over 50MW (though Welsh local authorities are responsible for considering applications to build power stations up to and including 50MW in Wales). Accordingly, it is envisaged that Wales would be included in the SSA process as set out in this Section.

## Justification

3.26  In common with certain industrial and domestic products, and some medical equipment, nuclear power stations involve the use of materials that give off ionising radiation. Although the radiation dose rates to workers and the public are much smaller than from natural sources of radiation, for example radiation from space or from rocks in the ground, they are subject to stringent European and UK regulations. Before the UK can adopt any new class or type of practice involving the use of ionising radiation, it must first be 'Justified', i.e. it must be demonstrated that any benefits resulting from its introduction outweigh the associated health detriment. We are all exposed to natural background radiation. Only 0.01% of the annual average dose of ionising radiation to a member of the public comes from the nuclear power industry. Over 80% of our annual radiation dose comes from natural sources (see Box 2).

3.27  In the UK, we already have a process in place to decide whether or not a new type or class of practice should be Justified. Additional guidance is being prepared setting out the process for the making and consideration of Justification application specifically in relation to new nuclear power stations. We consulted on this framework, both as part of the nuclear consultation and the supporting technical consultation[228]. An analysis of the responses to the technical consultation on the proposed Justification and Strategic Siting Assessment processes and the Government's response can be found at Annex B.

3.28  It is important that the Justification process runs as smoothly as possible and that there are no unnecessary delays in arriving at a Justification decision. In order to achieve this, we will put in place a project management framework to ensure that the timeframe is met and in a fair, transparent and robust way.

3.29  We are developing guidance, which will apply specifically to Justification applications relating to new nuclear power stations. This will set out in detail the process and timeframe for reaching a Justification decision in relation to the introduction of new nuclear power stations. A call for applications and guidance will be issued in February/March 2008.

3.30  The Justifying Authority takes decisions on Justification. In the UK, there are four Justifying Authorities; the UK Government in relation to England and reserved policy areas, and the three Devolved

228 www.berr.gov.uk/energy/whitepaper/consultations/nuclearpower2007/page39554.html

Administrations to the extent to which they have competence in respect of the subject matter of a particular Justification application. In the technical consultation[229] which accompanied the nuclear consultation we said since nuclear energy is a reserved matter, the sole Justifying Authority in the UK will be the Secretary of State[230]. We therefore expect the Secretary of State to be the sole Justifying Authority and expect that any Justification decision would be UK-wide. As the Justifying Authority, however, the BERR Secretary of State will, in accordance with the Justification of Practices Involving Radiation Regulations 2004, consult the Devolved Administrations. There is also a Concordat between the Government and the Devolved Administrations which sets out the working relations (including the setting up of a Justification Liaison Group) in a way which respects the devolution settlements.

## Generic Design Assessment

3.31   The nuclear consultation document set out plans to start a process of Generic Design Assessment (or pre-licensing). We also invited applications from vendors of nuclear reactor designs who were interested in having their design assessed through the GDA process. The consultation document set out that Generic Design Assessments would begin on a contingent basis while the consultation was ongoing. Following this invitation, the Joint Programme Office of the nuclear regulators (the Environment Agency and the Nuclear Installations Inspectorate and the Office of Civil Nuclear Security of the Health and Safety Executive) received applications from four vendors of nuclear reactor designs. These were:
*   AECL – for its ACR1000 design
*   Areva – for its EPR design
*   GE-Hitachi – for its ESBWR design
*   Toshiba-Westinghouse Electric Company – for its AP1000 design

3.32   The BERR website shows the letters of application and letters of endorsement from credible nuclear power operators[231]. On 5 July 2007, BERR announced that all four applicants had met the criteria set down in the consultation document[232]. The regulators subsequently agreed to assess all of the designs in the initial stage (phase 1) of the GDA. The initial stage of the assessment started in July 2007 and is expected to continue until spring 2008.

3.33   Generic Design Assessment is being led by the nuclear regulators to assess the safety, security and environmental impact of power station designs. The regulators involved are the Nuclear Installations Inspectorate (NII) and the Office of Civil Nuclear Security (OCNS) of the Health and Safety Executive (HSE), and the Environment Agency (EA).

229 The Future of Nuclear Power, *Consultations on the proposed processes for Justification and Strategic Siting Assessment*, URN 07/972, May 2007.
230 For civil nuclear power this is the Secretary of State for the Department for Business, Enterprise and Regulatory Reform.
231 http://www.berr.gov.uk/energy/whitepaper/consultations/nuclearpower2007/generic-design/page40336. html
232 GNN, *Four applications suitable for Nuclear pre-licensing*, 5 July 2007.

OCNS is a UK-wide regulator so the outputs from their Generic Design Assessment (GDA) work would be applicable across the UK. The NII, along with the rest of the HSE, regulates within Great Britain, so its findings from the GDA process in relation to heath and safety issues will apply in England, Wales and Scotland. The Health and Safety regulator in Northern Ireland (HSENI) has not been involved in the GDA process. Finally, the EA has a remit to regulate in England and Wales only. Environmental regulation is carried out by the Scottish Environment Protection Agency (SEPA) in Scotland and although SEPA were involved in the preparatory work for GDA, they have since withdrawn from the process. The body responsible for environmental regulation in Northern Ireland (The Department of the Environment) has not been involved in the GDA process, meaning that the EA's conclusions from the GDA process will apply in England and Wales only.

## Entering phase two of Generic Design Assessment

3.34    As set out in the consultation document, phase 2 of the Generic Design Assessment, which encompasses most of the detailed assessment work on the designs, is expected to run from early 2008 until 2010-2011. As this phase will be more demanding on the regulators' resources, it is unlikely that more than three designs can be assessed concurrently within the timeframe of 3 to 3½ years. This means that if all four designs that have been accepted for phase 1 of the assessment successfully proceed through this phase, we will need a further prioritisation process to select no more than three designs to proceed to phase 2.

3.35    As with the first prioritisation process – to decide which designs should enter phase 1 of Generic Design Assessment – the objective of this second stage of prioritisation will be to allow the regulators to focus their resources on those designs which are most capable of being licensed and operational in the UK within the 2016-2022 timeframe.

3.36    Before any second-stage prioritisation process takes place, the Government will write to the vendors of each of the reactor designs to ask them to confirm whether they would like their design to continue to be assessed in phase 2 (the detailed stage) of GDA, or whether they would prefer to defer their application until a future tranche of assessments. The Government will write in these terms to the vendors towards the end of phase 1 of the process and depending on their responses, this will determine whether or not a second-stage prioritisation process is needed. If all of the vendors want their design to continue to be considered for assessment, at the end of phase 1 we will instigate the prioritisation process we outline below.

## Second stage prioritisation process

3.37    We have identified three key considerations for this prioritisation process:
a)    Is the design likely to be deployable by 2016-2022?

    b)  Is this design likely to satisfy regulatory requirements without the need for significant modification?

    c)  What is the likelihood that this reactor will be deployed in the UK by 2016-2022?

3.38  These key considerations will enable us to ensure that the reactor designs chosen through the prioritisation process are those that investors are most likely to build in the UK within the 2016-2022 timeframe.

3.39  To meet consideration a), the Government will ask the vendors of the reactor designs to make a statement that their design would be deployable by 2016-2022 and to give their reasons for this assessment. The vendors will also be able to submit to the Government any other relevant information to support their case for prioritisation.

3.40  To meet consideration b), the Government will take account of information that emerges from phase 1 of the assessments, including the regulators' published reports and any other relevant factors the regulators may raise. The vendors will also be required to provide evidence that they will be able to provide full information in accordance with the requirements of the regulators, as laid down in the regulators' published guidance, in time for the detailed assessments to begin, or to show how they will provide any missing information in a timeframe that does not jeopardise the overall timetable. This should enable us to ensure that the timetable for the GDA process is not held up by having to consider designs that are not complete, or will not be complete in time to carry out the assessment within the required timetable. The vendors will also be required to provide evidence to show how they will address any initial regulatory concerns that emerge during phase 1 of the assessments.

3.41  To meet consideration c), the Government will canvas nuclear industry preference for designs, to ensure that those designs that have the greatest chance of being built by a future operator are prioritised. As in the first round of prioritisation that determined which designs would be considered in the initial phase of GDA, we will ask credible nuclear power operators to send letters of endorsement to the Secretary of State for Business, Enterprise and Regulatory Reform setting out which designs they wish to support for the detailed stages of GDA. The definition of a credible nuclear power operator is the same as for the first stage of the prioritisation and only those operators who meet the definition will qualify to endorse designs. This definition is set out below:

3.42  A credible nuclear power operator is one which:
- currently operates a nuclear power plant anywhere in the world; and
- currently operates an electricity generating station subject to UK health, safety and environmental regulation, or which has made a public commitment to become an operator of an electricity generating station (with a capacity in excess of 50MW) by 2016-2022 in a market subject to UK health, safety and environmental regulation. Note that the "in excess of 50MW" limit

**145**

is also used in regulation as the threshold for an electricity plant of significant size.

3.43 In this instance, as the regulators can assess up to three designs in the 3-3½ year timescale, operators will be able to nominate a maximum of three designs and must rank the designs according to their preference for deployment. To ensure transparency, we will ask operators to give reasons for their nominations and rankings. Operators should make reference to the criteria below in their response, although they will also be able to raise other relevant points.

## Criteria

- the level of advancement of a design
- their assessment of how easily the design could be built in the UK
- their willingness to invest in each design based on their view of its likely economic advantages, including the level of support they are giving each design throughout the GDA process
- their assessment of the likelihood that each design could be in operation by 2016-2020

3.44 The Secretary of State for BERR will consider all of the information provided by the vendors, the operators and the regulators, as well as any other relevant considerations that might become evident, before making a recommendation to the regulators, stating the three designs that the Government considers ought to be prioritised for the detailed stages of the assessment. We will put in the public domain our recommendation and the reasons behind it, including all relevant, non-classified information provided by the vendors, the credible nuclear power operators and the regulators. Based on this recommendation, as well as any other factors they consider relevant, the regulators will decide where to deploy their resources and will inform the vendors of their decision.

3.45 This prioritisation process will take place when the regulators have completed their assessments following phase 1 of the GDA, which is estimated to be around Spring 2008. At the end of phase 1, the Government will inform the vendors and will announce the start of the prioritisation process. From that point, the vendors and operators will have a short period to provide the relevant information to the Government. In due course, the Government will make an announcement to confirm the dates on which this period will begin and end and will provide details of where to send the information. The Government's recommendation will be communicated to the regulators and made public as soon as it has completed analysis of the information received.

# Waste and decommissioning regulation

## *Paying for waste and decommissioning costs*

3.46 It is the Government's policy that the owners and operators of new nuclear power stations must set aside funds over the operating life of the power station to cover the full costs of decommissioning and their full share of waste management and disposal costs. Waste management costs include the costs of final disposal of the waste, as well as interim storage and any treatment and packaging. These financing arrangements must be robust, and designed to deliver sufficient funds to cover these costs in a number of different scenarios.

3.47 The Secretary of State for Trade and Industry (now Secretary of State for Business, Enterprise and Regulatory Reform) appointed Dr Tim Stone, a senior financier with experience of major capital investment projects, in January 2007, to advise Government on financing the costs of decommissioning and waste management and disposal costs for new nuclear power stations. Dr Stone reports to the Secretary of State for Business, Enterprise and Regulatory Reform and the Chief Secretary to the Treasury.

3.48 In our nuclear consultation published in May 2007, the Government set out three options for funds to hold monies for decommissioning and waste management:
- the nuclear operator could accumulate funds in a ring-fenced way within the company itself, or
- the nuclear operator could make specified payments direct to Government, or
- the operator could make payments to a separate, independent fund, such as a trust

3.49 Government has determined that independent funds, outside of the control of nuclear operators, should be created to accumulate and manage payments from the operator to meet the full costs of decommissioning and full share of waste management costs. This approach would be transparent and would be consistent with the policy of ensuring that operators, not Government, take full responsibility for meeting the costs of decommissioning and waste management. These independent funds would be insulated against the commercial fortunes of the operator, would be invested prudently and could be liquidated reasonably readily as required to discharge the liabilities stemming from waste and decommissioning.

### Setting a framework through the Energy Bill

3.50 We expect to introduce the Energy Bill to Parliament in January 2008. Government intends to ensure that the operators of new nuclear power stations meet the full costs of decommissioning and their full share of waste disposal costs. It will do this by imposing new legal duties on operators in this regard and creating new powers for Ministers to ensure that operators meet those duties under the Energy Bill.

**147**

3.51 As introduced to Parliament, the clauses in the Energy Bill:
- require operators of any new nuclear power stations to submit a funded decommissioning programme for approval by the Secretary of State for Business, Enterprise and Regulatory Reform. The funded decommissioning programme must set out:
  - the steps operators will take to decommission the installation, clean up the site and manage waste (including spent fuel) produced during its electricity generating life
  - the estimated costs of taking these steps
  - how operators intend to meet those costs, and
  - details of the financial security to be put in place to meet the costs identified
- Give Ministers a power to approve the funded decommissioning programme, approve it subject to modifications or conditions or reject the programme
- Impose a duty on operators to comply with the programme by making it an offence if they fail to do so
- Give Ministers powers:
  - To require information from the operator, any persons responsible for the fund, and any other persons with obligations under the programme to find out whether they are complying with the programme
  - Where the operator, the persons responsible for the fund or any other persons with obligations under the programme are not complying with the programme
    - to obtain information from other bodies corporate "associated" with the operator (to enable Ministers to consider whether to impose obligations on such persons)
    - to direct persons in breach to take the action necessary to bring themselves back into compliance
- Enable Ministers to require operators and persons responsible for the fund to carry out regular reviews of the funded decommissioning programme[233]
- Give Ministers powers to approve modifications to the funded decommissioning programme that might be proposed by the nuclear operator from time to time and, in certain circumstances, to require modifications.

3.52 Our policy on waste and decommissioning for new nuclear power stations is designed to ensure that operators make adequate arrangements to cover the full costs of decommissioning and a full share of waste management costs. Operators are responsible for decommissioning and waste management costs. If the protections we are putting in place through the Energy Bill prove insufficient, in extreme circumstances the Government may be called upon to meet the costs of ensuring the protection of the public and the environment.

---

233 We intend to create a new independent advisory body, the Nuclear Liabilities Financing Assurance Board to provide scrutiny and advice on the suitability of decommissioning programmes – see Box 4.

3.53    The Bill seeks to ensure that for waste management liabilities which arise during the station's electricity generating life, there is flexibility in terms of what will be regulated for financial purposes. This is because it may be sufficient to permit operators to pay for some of these costs from their revenue (for example).

3.54    The new provisions included in the Energy Bill will apply to England, Wales and Northern Ireland. If there is a change in policy towards new nuclear power stations in Scotland we would seek to extend the provisions in the Energy Bill to Scotland at the earliest available opportunity.

## Guidance on approvable arrangements

3.55    In parallel with the Energy Bill, the Government will publish for public consultation two sets of draft guidance on what an approvable funded decommissioning programme should contain. This guidance will assist businesses in understanding their obligations under the Bill. The first set of guidance will assist businesses in setting out and costing the steps involved in decommissioning a plant and managing radioactive waste and spent fuel in a way which Ministers may approve. The second set of guidance will assist operators in setting out acceptable proposals for how sufficient funds will be accumulated to meet the costs identified. The guidance will state that nuclear operators should establish a Fund or Funds to which they will make regular payments to accumulate monies to cover the costs of decommissioning and waste management. The Government believes that a fund is the best way to achieve its objectives and satisfy the principles set out below. This second set of guidance will set out the guiding principles against which Government will assess the funding proposals submitted by nuclear operators for approval under the Energy Bill. We describe the proposed principles below.

## Sufficiency of funds

3.56    The contribution schedule to and investment strategy of the Fund or Funds should be designed and followed to deliver as and when needed sufficient funds to discharge in full the operator's liabilities for decommissioning and waste management and disposal.

3.57    Operators would be required to ensure that they have adequate financial safeguard mechanisms in place to top up an insufficient Fund, for example where:
- the power station has to be closed and decommissioned early for technical reasons, or
- the operator becomes insolvent before the station has reached the end of its electricity generating life and no suitable buyer can be found for the station who is willing to meet the operator's liabilities, or
- during decommissioning, the fund proves inadequate to meet the operator's liabilities.

**149**

3.58 The funding arrangements should ensure that the prospect of the operator's liabilities having to be met in whole or in part from public funds is remote at all times.

## Independence of funds

3.59 The arrangements relating to the accumulation, management and disbursal of monies necessary to discharge the operator's liabilities should be set out by the operator and approved by the Government but subsequently overseen independently of both the operator and of the Government. It should not be possible for the Government or the operator to control funds.

## Restrictions on use of funds

3.60 The arrangements the operator puts in place should ensure that funds may not be used for any purpose other than decommissioning and waste management and disposal even in the event of the failure or insolvency or reorganisation of the operator.

## Transparency

3.61 The funding arrangements should ensure that the process of accumulating and maintaining and protecting funds sufficient to discharge the operator's liabilities is clear to Ministers, operators and the public. There would need to be transparency and separate reporting of the two sets of liabilities (decommissioning and waste disposal) and the monies available to meet the costs of each and there should be no element or prospect of cross-subsidy between the two.

3.62 The consultation on the guidance will set out the principles listed above and will enable the Government to test its view that to meet the principles each operator's fund should have the following characteristics:
- it should be a separate entity from the operator
- it should be administered by a group of people the majority of whom are independent of the operator
- it should receive payments from the operator
- it should be responsible for managing and investing monies it receives
- it should generate a sum of money to meet the operator's liabilities before the station reaches the end of its generating life.

3.63 In addition, the consultation on guidance will set out for comment the information that the Government proposes to offer industry in terms of fund structure; fund governance; the process for review of the fund's performance and cost estimates; investment strategy for the fund; how monies should be disbursed from the fund; change control of the operator; winding up the fund; and protection to top up an insufficient fund.

## Cost estimates and a Base Case for decommissioning and waste management

3.64 To ensure that the Government can have confidence that owners/ operators of any new nuclear power stations make contributions that meet the full costs of decommissioning and their full share of waste management costs, it will be important to understand the likely costs of these activities. As the nuclear consultation document announced, the Government has embarked on a programme to determine robust estimates of costs.

### Base Case

3.65 To enable us to estimate the potential costs of waste management and decommissioning and to ensure adequate provision for their financing, we are working to set out a means for waste management and decommissioning that will be costed. We will call this the "Base Case". It will build on existing policy and regulations for waste management and decommissioning; it will also make additional assumptions to ensure that it represents a realistic and prudent way to estimate the costs of these activities. In particular, we intend the Base Case to set down among other things:

- the need for operators to provide interim storage facilities, capable of being maintained or replaced to contain waste for an extended period of time until we expect a geological disposal facility to be in a position to accept waste from new nuclear power stations and beyond that date to provide some contingency
- the treatment and disposal of low-level waste
- how soon decommissioning would take place after closure
- when and on what terms we would assume that waste could be transferred to a geological disposal facility.

3.66 Operators of any new nuclear power stations will need to have regard to the provisions in the Base Case when developing the programme they will submit to Government, although there will be flexibility to allow companies to propose more effective ways of dealing with decommissioning and waste management if they choose to do so. This is because the Base Case is likely to take effect as guidance issued under the provisions of the Energy Bill.

3.67 As well as meeting current regulatory requirements, each operator's programme will be subject to Government approval to ensure that it includes all the elements for which operators will need to make financial provision. Once a programme is approved, the operator will be required to follow it. They will, however, be allowed to amend the programme, subject to Government approval. We will require operators to update the programme to reflect modifications such as operational or technical changes to a nuclear power station that would have an impact on the estimates of decommissioning costs.

**151**

3.68 We are working with the regulators, the NDA and key stakeholders to develop the Base Case. We plan to launch a formal consultation in early 2008 on what an approvable decommissioning programme should contain.

## Cost estimates

3.69 The Base Case is a key input into our work to develop robust estimates of the costs of waste management and decommissioning for new nuclear power stations. To provide further inputs, we have carried out an exercise to develop our understanding of the waste inventories that would be produced by different generic reactor types, to determine the volume and types of waste that new nuclear power stations could produce. To derive estimates of the costs of waste management and decommissioning for new nuclear power stations we are developing a cost model that will enable us to produce a range of likely costs, as well as giving us information on the level of certainty of those costs. The consultation we plan for early 2008 will include further information on our cost modelling work. Alongside the consultation, we will publish a roadmap that sets out a timeline to publishing cost estimates.

## Full share of waste management costs

3.70 The Government has stated as policy that operators of new nuclear power stations will be obliged to meet their full share of waste management costs.

3.71 We are modelling the financial impact of adding waste from new nuclear power stations to a repository that would otherwise only be designed to hold only the UK's existing waste inventory. Our modelling will take into account the additional direct costs, for example through needing to construct additional underground caverns to accept waste from new nuclear power stations. We will then consider which other items of cost the addition of waste from new nuclear power stations might affect less directly. In this way, we will be in a position to ensure that the price that operators pay for disposal of their higher-activity wastes in the government-provided geological disposal facility reflects their full share of the costs of adding waste from new nuclear power stations to this facility. These costs will include a proportion of the fixed costs of building a geological disposal facility. It may be more difficult to quantify these indirect costs and our methodology for defining the "full share of costs" will address this issue in determining the contribution that operators of new nuclear power stations ought to make. The Government will, early in 2008, set out further detail alongside the consultation on what an approvable decommissioning programme should contain.

## Clarity on the costs for disposal of waste from new nuclear in a Government facility

3.72 Potential investors in new nuclear power stations need clarity on the maximum amount that they would be expected to pay for the Government to take responsibility for their future waste in a geological

disposal facility. This cost certainty would enable them to take investment decisions and seek financing.

3.73  In response to the consultation, energy companies have indicated that they would be prepared to pay a significant risk premium, over and above the expected costs of disposing of waste, in return for having the certainty of a fixed upper price.

3.74  The Government plans to use the exercise on waste cost modelling to set a fixed price or upper limit for nuclear operators. This price would be set at a high level, including a material risk premium over and above expected costs. This risk premium will help to ensure that the operator bears the risks around uncertainty in waste costs and will provide the taxpayer with material protection. Should the actual costs of providing the service prove lower than expected, these lower costs will not be passed on to nuclear operators, who would have gained from certainty of a fixed price and would not have been exposed to the risk of price escalation.

3.75  We will publish further information on pricing waste disposal in early 2008 alongside the consultation on decommissioning programmes. Proposals will be subject to ensuring compliance with EU State Aid law.

## Nuclear Liabilities Financing Assurance Board

3.76  Details of the new Nuclear Liabilities Financing Assurance Board are set out in Box 4 and in Annex C.

**BOX 4: NUCLEAR LIABILITIES FINANCING ASSURANCE BOARD**

- In recognition of concerns raised in the consultation, we intend to create a new independent advisory body, the Nuclear Liabilities Financing Assurance Board (NLFAB). This new board will provide independent scrutiny and advice on the suitability of the decommissioning programmes submitted by operators of new nuclear power stations.
- The NLFAB will advise the Secretary of State for Business, Enterprise and Regulatory Reform on the financial arrangements that operators plan to put in place to cover waste management and decommissioning. The NLFAB will also provide advice to the Secretary of State on the regular reviews and ongoing scrutiny of funding arrangements, once new nuclear power stations are operational.
- The Board is expected to consist of experts from relevant fields such as current or former fund managers, pension trustees, actuaries and nuclear engineers. The board members will be appointed by the Secretary of State.
- The NLFAB will be a purely advisory body and will have a tightly defined scope focused solely on ensuring that the outcomes intended will be delivered and that robust financial arrangements for decommissioning and waste management disposal are put in place by operators.

## Tax

3.77 It is not intended that incentives will be provided through the fiscal regime to invest in nuclear power generation in preference to other types of electricity generation. The Treasury and HMRC are, however, exploring the possibility that the timing of nuclear decommissioning could create a potential tax disadvantage for nuclear operators and, if so, whether it may be appropriate to take action to ensure a level fiscal playing field between nuclear power and other forms of electricity generation.

## Scope

3.78 We recognise that in due course energy companies may come forward with proposals to develop other nuclear installations and facilities that will both sustain and support the development of a growing nuclear energy sector. Should the sector develop in such a way, Government would seek to ensure that such developers of installations or facilities which are constructed for a purpose connected to the generation of electricity by nuclear power stations cover their full decommissioning costs and full share of waste management costs. As introduced to Parliament the Energy Bill gives Ministers a power to extend the clauses in the Bill to such installations so as to ensure this objective is met.

# Alternatives to nuclear power

**This annex looks at alternative futures for achieving our long-term energy goals. It uses updated analysis originally conducted for the Energy White Paper and the Nuclear Consultation Document to outline the implications for our energy goals if we ruled out the option of new nuclear power stations.**

## Introduction

A1    Our two key energy challenges are to tackle climate change by reducing carbon dioxide emissions and to ensure secure, clean and affordable energy as we become increasingly dependent on imported fuel.

A2    There is great uncertainty about future energy demand, the pace of technological change and the future availability and cost of energy supplies. In this context, the Government's view is that the policy environment should provide a framework that allows investors to consider a portfolio of different technology options that are consistent with our goals to reduce carbon dioxide emissions and to achieve security of supply. Reliance on single solutions on their own will not allow us to meet our goals under all circumstances. By having diverse options, the UK will be better placed to deal with the range of possible futures that could unfold.

A3    This annex sets out our analysis of how the UK could deliver its goals for energy policy without new nuclear power stations and looks at the implications for our carbon emissions reductions targets, security of supply and costs. The analysis focuses on the implications for the medium to long-term energy mix since, given the lead times for new nuclear power stations, this is the period when new nuclear power stations could reasonably be expected to make a contribution to delivering Government's energy policy goals. It shows that to achieve our goals, without giving energy companies the option to build new nuclear power stations, would imply:

- Further need for significant improvements in energy efficiency across all sectors beyond what is set to be achieved through current policy
- Higher cost emissions reductions in the electricity generation sector and greater effort to reduce emissions through more costly options outside the electricity generation sector, for example transport
- A higher risk of not significantly reducing carbon emissions in the electricity generation sector, because of greater reliance on

**155**

low-carbon electricity generation technologies which are subject to greater risks in terms of their feasibility and deployability, e.g. wind and carbon capture and storage (CCS). This could also have implications for security of electricity supplies

- A mix of electricity generation technologies with less diverse characteristics than if new nuclear power stations were to be an option, with implications for ensuring energy security under the widest range of future scenarios.

A4   The Government's view is therefore that excluding nuclear power as an option would make it more challenging and expensive to meet our goal to reduce carbon emissions and could expose the UK to greater security of supply risks because our electricity supplies would be less diverse.

# Carbon emissions

## Contribution of nuclear power to reducing carbon emissions

A5   The Climate Change Bill[234] commits the Government to legally binding targets to reduce carbon emissions by at least 60% from 1990 levels by 2050[235]. This is equivalent to a fall in annual carbon emissions of around 317 $MtCO_2$ from 2005 levels[236]. All sectors of the economy will need to contribute if we are to achieve such an ambitious target.

A6   The electricity sector is currently responsible for about a third of the UK's total $CO_2$ emissions, emitting around 172 $MtCO_2$ in 2005. To achieve our 2050 target at minimum cost, the electricity sector will need, over the long-term, to considerably, if not fully, 'decarbonise', since reducing emissions from the electricity sector is in general relatively less expensive than in some other sectors, such as transport. However, in the medium to longer term, the King review of low carbon cars concluded that electric or hydrogen-powered vehicles will contribute to decarbonisation of road transport. This will require major technological breakthroughs as well as substantial progress towards decarbonising the power sector[237]. This could be achieved on the demand side, through action to save energy, and on the supply side, through the development of low-carbon generation technologies, such as renewables, CCS and nuclear power.

234 http://www.defra.gov.uk/environment/climatechange/uk/legislation/index.htm
235 The evidence now suggests that as part of an international agreement developed countries may have to reduce their emissions by up to 80%. This evidence will be considered by the Committee on Climate Change, which will advise Government on whether our own domestic target should be tightened up to 80%.
236 Emissions in 2005 were 553 Mt $CO_2$, compared to a target for 60% by 2050 of 236Mt $CO_2$.
237 http://www.hm-treasury.gov.uk/media/9/5/pbr_csr07_king840.pdf

A7    Nuclear power now provides approximately 19% of our electricity generation[238], 7.5% of total UK energy supplies[239] and 3.5% of total UK energy use[240]. As a low-carbon source of electricity, nuclear power makes an important contribution to lowering the carbon intensity of our energy supplies. Without our existing nuclear power stations, the UK's $CO_2$ emissions would currently be some 29 to 59 $MtCO_2$[241] higher than otherwise. However, most of the existing nuclear power stations are due to close in the next 15 years or so, based on published lifetimes. Over the same period, 12 GW of coal and oil-fired power plants are due to close as they reach the end of their lifetime and due to EU environmental legislation.

A8    It will be for energy companies, taking account of the changes in energy and electricity use and the Government's policy framework, to decide on the type of capacity that will replace the power plants that close over the next 20 years and beyond. It is therefore difficult to predict how the energy system and the electricity mix will develop in the long-term, that is over the next 40 to 50 years. Investors face great uncertainties in planning future power station projects. For example, on top of uncertainties about likely future demand for electricity, it is very difficult to predict the cost and availability of fossil fuels, or the cost and feasibility of existing and emerging low-carbon technologies.

A9    To understand the long-term implications of allowing or not allowing energy companies to invest in new nuclear power stations, we have used analysis conducted for the Energy White Paper 2007, the Nuclear Consultation document and the Climate Change Bill Impact Assessment. This analysis used a model of the UK energy system, the Markal-macro model, to analyse different long-term scenarios and how these affect the combination of technologies that could allow us to achieve our 2050 target for $CO_2$ emissions at least cost (see Box A1)[242].

A10   The Markal-macro model indicates that, under a range of different assumptions, a diverse and very low-carbon electricity generation mix would be the best way to reduce $CO_2$ emissions across the economy as a whole[243].

238 The May 2007 consultation document stated that nuclear power accounted for around 18% of electricity, based on the latest energy statistics available at that time. The most recent published data now available, in the Digest of United Kingdom Energy Statistics 2007, shows that in 2006 nuclear power accounted for 19% of the electricity generated in the UK.
239 This figure is the total amount of fuel used to generate electricity taken as part of total energy supplies. This issue is discussed in Section 2 of this White Paper.
240 See the simplified flow diagram of UK energy supply and consumption 2006 showing the role of nuclear at http://www.berr.gov.uk/files/file43008.pdf.
241 Depending on the assumption made about the alternative capacity to replace nuclear.
242 For more details and the report on the work carried out for the Energy White Paper 2007 see http://www.ukerc.ac.uk/TheMeetingPlace/Activities/Activities2007/0706MARKALMacroEWP.aspx.
243 Full details of the assumptions and the methodology used in the Markal model are available on the UK Energy Research Council (UKERC) website http://www.ukerc.ac.uk/ResearchProgrammes/EnergySystemsandModelling/ESMMARKALModelDocs.aspx.

## BOX A1 MODELLING APPROACH TO DEVELOP LONG-TERM SCENARIOS FOR THE ENERGY SECTOR

The UK MARKAL-macro model

MARKAL stands for MARKet ALlocation, since it mimics a market by always choosing the combination of technologies with the lowest cost. The MARKAL-macro model can be linked to a simple economic growth model, which represents the relationship between carbon, energy prices and energy demand. The combined MARKAL-macro (M-M) model gives estimates of future GDP, as well as the costs of carbon abatement in terms of a proportion of GDP. The M-M model covers the entire energy system, including electricity, heat and transport and is one of the few models that can explore the energy system in the long-term. We used the M-M model to explore different scenarios for the mix of technologies used to generate electricity in 2050, all of them consistent with achieving our goal of reducing $CO_2$ emissions by 60% by 2050 in the cheapest way. We conducted this analysis for a number of scenarios in order to capture the range of impacts on costs that might materialise under different assumptions about fuel prices and technology costs. The central scenarios used cost assumptions collected by the UK Energy Research Centre from a wide array of published sources and peer reviewed by a panel of experts. Other scenarios were developed using different ('DTI') assumptions based on numerous published market studies, reflecting estimates for typical projects being developed in the UK and published by DTI with the Energy Review Report, July 2006.

The MARKAL-MACRO model has both strengths and weaknesses. While its assumptions on data, technology pathways and constraints are transparent, not all factors can be captured fully. By optimising costs, in effect it represents a perfect energy market, and neglects barriers and other non-economic criteria that affect decisions. It also assumes that there is perfect foresight about the development of technologies and their costs so that at any point in time the model knows when and at what costs different technologies will become available. In addition, the model therefore does not capture the full range of uncertainty around the development of new technologies. In this way, it underestimates the full costs and risks of bringing less-developed technologies to market and the impact this could have on delivering our carbon goals while ensuring security of supply and affordability.

A11　Our current approach to estimating savings in $CO_2$ emissions from Government policy measures is to assume that any new low-carbon generation plant that is built displaces new fossil fuel power stations, therefore saving the $CO_2$ emissions equivalent to that type of plant[244]. This is also the assumption made in the Markal model[245]. According to the model, nuclear power could, in 2050, deliver carbon emissions

244 The Nuclear Cost Benefit Analysis (http://www.berr.gov.uk/files/file39525.pdf) assumed that the alternative type of generation that would be displaced by new nuclear power would be gas-fired plants.
245 Under the base case and without any carbon emissions constraints, in the Markal model coal fired plants would be the preferred type of generation.

savings of up to 97 $MtCO_2$. If new nuclear power stations were not allowed, such savings would need to come from other options.

## Achieving our $CO_2$ targets without new nuclear power stations

A12 We have analysed scenarios in which the Markal model is prevented from allowing new nuclear power stations to be built, while still enforcing the requirement to meet the UK's goal of reducing carbon emissions by 60% by 2050, and to do so at least cost to the UK economy (see Box A2). If there is no option to build new nuclear power stations, the model shows that, to meet our target, there would have to be even more effort to:

- Improve energy efficiency across all sectors
- Further invest to deploy alternative low-carbon electricity generation technologies, such as wind power and carbon capture and storage, and
- Further reduce carbon emissions in other sectors of the economy, for example transport.

## BOX A2 THE ELECTRICITY MIX IN 2050 – A MODELLING PERSPECTIVE

Most of the scenarios explored in the work with the Markal model included nuclear power as an option, while some explicitly excluded the option of investing in new nuclear power stations. It also covered scenarios on different economic growth rates, fossil fuel prices and technological developments. Significant improvements in energy efficiency are common across all scenarios. To meet our carbon reduction goals, energy demand would need to be lower than today despite the economy being much larger (UK GDP is expected to grow from around 1.2 trillion pounds today to around 2.8 trillion in 2050 in real terms). If new nuclear power stations were allowed as an option for investors, under all scenarios the model shows that new nuclear power stations would be part of the mix of technologies that delivers our carbon goal at least cost to the economy[246].

In the absence of nuclear power, according to the model renewable energy sources would account for around 41% of the 2050 electricity generation mix, with coal with CCS also making up 41%. Most of the renewables would come from wind technology, both onshore and offshore, with the remaining electricity generated through biomass or hydro. Some gas-fired power stations would still remain in the mix, mainly to provide flexibility to the system. Excluding CCS would result in the electricity mix being dominated by offshore wind power (around 60% of generation), supplemented by higher cost renewables, including marine (bringing the total share of renewables to over 80% of the mix), with natural gas and bio-gas CCGT plants meeting requirements for seasonal minimum electricity demand.

As part of further work for the Climate Change Bill, Defra commissioned a further study using the MARKAL-Macro model to consider the additional impacts (economic and technological) of reducing UK $CO_2$ emissions by 70 per cent and 80 per cent by 2050, beyond the current UK goal of a 60% reduction[247]. Similarly to the previous work, the model was tested under different scenarios, some of them explicitly excluding the option of investing in new nuclear power stations. This analysis shows that increasingly stringent carbon constraints force even more radical change on the energy system, in terms of energy mix and technology take-up, than seen in the previous 60% constraint runs. For example, limits on new nuclear power stations result in an increase in wind generation to meet demand beyond 60% of the mix, but also in the take up of further expensive abatement measures. The marginal abatement cost in 2050 could also increase from around £800/tC in the base case to around £1450/tC when both nuclear and CCS are not available. An 80% reduction scenario using central fossil fuel assumptions with the option of new nuclear power stations would reduce GDP in 2050 by 1.6%. Placing a constraint on new nuclear power stations would increase the costs to 1.7% of GDP in 2050.

---

246 See UKERC, *Final Report on DTI-Defra Scenarios and Sensitivities using the UK Markal and Markal-Macro Energy System Models*, May 2007.
247 *MARKAL Macro analysis of long run costs of climate change mitigation targets*, November 2007.

## *Energy efficiency*

A13   According to the results from the UK Markal-macro model, by 2050, with no carbon constraint, electricity demand is expected to increase by around 30% compared to today's levels. Energy efficiency can make a significant contribution to achieving our policy goals cost effectively. When all options, including new nuclear power stations, are available, the model shows that improvements in the efficiency and the way we use energy could reduce demand by around 30% compared to what it would otherwise be. Total electricity demand would therefore remain at roughly today's levels despite the UK's GDP being three times larger than it is today.

A14   If new nuclear power stations were to be excluded from the generation mix, efforts to improve energy efficiency would need to go even further. The UK Markal-macro model shows that electricity demand would need to fall by around 6% compared to today's levels to meet our long-term carbon reduction target in this scenario. Such a reduction, which is equivalent to all the electricity currently consumed in private offices, would have to occur in the context of continued economic growth. A substantial additional improvement in the energy intensity of the economy would therefore be required. In other words, according to the model we would need to produce each unit of economic output with less and less energy, around 48 tonnes of oil equivalent per million pounds compared to 211 tonnes of oil equivalent per million pounds today[248].

A15   Without such improvements in the energy efficiency and intensity of our economy, the only other way to keep energy demand at current levels or even to reduce it, would be to forego economic growth. Such an option is not directly considered in the model but would imply a considerable change in consumer behaviour, with impacts on standards of living.

A16   Energy efficiency measures (such as improved billing for businesses and improved insulation in homes) are amongst the most cost effective ways of reducing energy demand and hence carbon emissions (as illustrated by the Marginal Abatement Curve in Section 2 of this White Paper). However, evidence also points to the fact that despite the benefits most consumers fail to take up the opportunities available to them, even if policy can be effectively designed to overcome these. There are limitations to the potential contribution that energy efficiency measures can make in delivering our energy goals. For example, households may use the financial rewards from improving energy efficiency to increase their use of energy – a phenomenon described as the "rebound effect" – either to improve their level of comfort, by increasing the temperature at which they heat their home, or by purchasing more energy-consuming products, thereby increasing carbon emissions. So we need to factor in the overall impact on

248 When nuclear is available as an option the ratio would be 61 tonnes of oil equivalent per million pounds.

A White Paper on Nuclear Power

reducing carbon emissions that result from action to improve energy efficiency[249].

A17  Therefore, even if we achieve the reductions in future electricity use as estimated in the UK Markal-macro model, the UK is still likely to need a substantial amount of new electricity generation capacity in the coming decades, which will need to be low-carbon if we are to meet our climate change goals.

## Generation sector

A18  To almost fully decarbonise our electricity supply will require us to replace all existing generation capacity with low-carbon technologies by 2050. There are a number of low-carbon technologies that energy companies could invest in, some are already available, e.g. nuclear or wind power; others are in development, the primary example being CCS. To meet the UK's carbon reduction goal, the Markal-macro model shows a diverse generation mix in 2050, with nuclear between 5% and around 60% of the electricity mix by 2050 (depending on assumptions), and renewables and CCS making up most of the remaining capacity.

A19  When nuclear power is excluded from the electricity generation mix, in order to reduce emissions from the generation sector and meet our carbon emissions goals, more investment would need to go towards other low-carbon generation technologies. The total contribution of renewable technologies and CCS would have to increase substantially in such a scenario (see Box A2). This is even more significant if we were to have higher targets for emissions reductions in 2050[250].

A20  According to the Markal Macro model, when new nuclear power stations are excluded, electricity generated from renewable sources would have to play a significant role in electricity generation, constituting over 40% of the generation mix by 2050. The proportion of supplies generated through wind power would need to rise from around 1 to 2% of electricity output today to around 30%. In the context of the EU 2020 renewables target, we will next year set out a renewable energy strategy to significantly increase the proportion of UK electricity generated from renewables by 2020. Achieving this will mean resolving the challenges associated with higher penetrations of renewables.

A21  Beyond 2020, however, further cost effective reductions in emissions from the electricity generation sector will require a complementary effort to deploy non-renewable low-carbon technologies. Without new nuclear power stations, the main low-carbon non-renewable electricity generation technology which could be deployed would be CCS, but this will only be possible if this technology can be successfully demonstrated in the next decade. The runs in the Markal-macro model which excluded new nuclear power stations show coal plants with CCS

249 The UK Energy Research Council estimates that rebound effects in the energy market can be significant when both direct and indirect consequences are taken into account.
250 Our target is to put ourselves on a path to cutting the UK's carbon dioxide emissions by at least 60% by about 2050, with real progress by 2020. There is provision in the Climate Change Bill for the targets to be amended in light of significant developments in climate science or in international law or policy.

making a significant contribution to reducing emissions in the electricity generation sector, providing up to 41% of electricity supplied, which is roughly equivalent to the amount of electricity provided today by all gas-fired plants.

A22   However, large-scale CCS represents a significant technological challenge. No commercial scale power station using CCS technology has yet been developed anywhere in the world, although all the elements of the individual stages of the process have been demonstrated. Given the huge potential of CCS to abate carbon emissions in the UK and abroad, a number of governments, including the UK, are in the process of supporting commercial scale demonstrators of the technology on power generation[251].

A23   To reflect the uncertainties over CCS, we have, therefore, also examined a scenario where new nuclear power is excluded as an option, and safe, reliable CCS with power generation fails to develop on a sufficient scale in the timeframe available. This scenario could, for example, result if the overall costs of a large-scale roll-out of CCS, including the required infrastructure, turn out to be substantially higher than projected. Such a scenario would require us to further reduce the amount of energy we use by 2050 and substantially affect the means of producing it:
- Electricity demand would have to fall by around 9% relative to today's levels, compared to essentially no change in demand if new nuclear power stations and CCS were included in the mix. Again, this would need to be delivered against a background of the UK economy in 2050 being almost three times as large as today
- Renewables generation would have to provide up to 80% of electricity supplies, with wind generation providing around 60% of the UK's electricity generation mix. This would require further renewables deployment beyond our commitment to the EU 2020 renewables target, which would bring further challenges in terms of finding appropriate locations, grid connections and to overcome local objections
- Further increasing generation capacity from intermittent renewable sources such as wind would require additional investment in conventional generation plant to provide 'back-up' for the inevitable hours when intermittent renewable energy resources would not be available[252].

251 Currently three governments worldwide are supporting commercial scale demonstrations plants.
252 The Markal model does take into account the need for back-up generation capacity, but its representation is simplified and therefore does not capture the full implications of changes in the system needed to accommodate intermittent generation technologies.

## BOX A3 ADDITIONAL EFFORT FROM OTHER SECTORS WHEN NEW NUCLEAR POWER IS NOT AVAILABLE

Analysis from the UK Markal-Macro model showed that when nuclear power is excluded as an option, minimising the costs of meeting the 2050 carbon reduction goal would require other non-electricity sectors to further reduce carbon emissions. By excluding new nuclear power stations and therefore reducing the low-carbon options in the electricity generation sector, carbon emissions from electricity generation are likely to be higher than under the scenario where new nuclear power stations are allowed.

Excluding the option of new nuclear power stations could have broader energy policy implications, as we would have to consider options that are more expensive than new nuclear power stations both within and outside the electricity generation sector (see Marginal Abatement Curve in chapter 10 of the Energy White Paper and Section 2 of this White Paper). In other words, we would have to make even greater reductions in emissions from transport and the use of heat. For example, when we exclude nuclear power from the mix, the model shows that $CO_2$ emissions from the transport sector will need to be around 10% lower. This would have to come primarily through demand reduction and increased efficiency, thus reducing car fuel demand by approximately 13%. Some increase in hydrogen use in transport might also be required.

A24   The changes we could see in electricity generation are not limited to conventional generation. Distributed Generation (DG) – generation of electricity and production of heat close to its point of use – could play a significant part in our future energy mix. Currently around 5% of our electricity comes from DG, primarily through gas-fired Combined Heat and Power (CHP) plants. The Markal model under the different scenarios estimates that up to 25% of our electricity could come from DG in the future, mostly from CHP but also from newer technologies including microgeneration. As with other low-carbon options, the penetration of Distributed Generation in the future mix will depend on the level of ambition in reducing carbon emissions and on the availability of other technologies.

A25   The Energy White Paper made it clear that the Government sees potential advantages from more use of DG alongside the traditional centralised system. DG can contribute to meeting carbon emission reduction targets in a variety of ways: making use of the waste heat produced through electricity generation to heat and cool buildings; reducing electricity losses by moving generation much closer to where electricity is used; reducing the need for transmission and distribution infrastructure; facilitating the use of local renewable energy sources; and encouraging behavioural change through increased awareness of energy consumption[253].

253 Energy White Paper, *Meeting the Energy Challenge*, Chapter 3, URN 07/1006, May 2007.

A26 Preliminary findings of the analysis conducted for the Energy White Paper suggest that the costs of some DG technologies may be competitive with costs of centralised technologies. However, overall systems costs and risks to our energy security are likely to be lower if we retain a framework where DG is a complement to rather than an alternative to centralised electricity generation.

A27 We recognise that there is a large degree of centralisation in the UK energy system and that this may disadvantage smaller players. More use of DG will bring new and smaller players into contact with the system and it is important that the costs and complexities of this interaction are not prohibitive. The Energy White Paper committed us to level the playing field, tackling the barriers to help more distributed solutions come to market and become cost-competitive. On 18 December 2007 we published a joint consultation with Ofgem[254] which aims to make the regulatory arrangements "fit-for-purpose" for DG, reducing costs and burdens and making it significantly easier for projects to work with the regulated electricity system, in terms of paying a fair price for use of the network, being rewarded for the benefits they can bring and reducing the risks of participating in the electricity markets.

A28 The scale of the challenge to achieve a higher penetration of DG in the future generation mix in a cost-effective way is nevertheless significant but we believe the changes to the policy framework we have set out will make DG a significant part of our future energy mix. However, the pull through and benefits of each policy cannot and should not be taken in isolation, or assumed to mean that DG would develop such that it would replace the need for conventional centralised sources of energy.

A29 Therefore, as also demonstrated by results from the Markal model, we believe that in the future there will be continued need for a considerable amount of centralised generation capacity, and that this will need to be low-carbon and cost-effective. In this context, we believe that new nuclear power stations would not constrain the potential from DG.

## Security of supply

A30 In the absence of nuclear power, as outlined above, the UK generation mix would need to be more reliant on other forms of low-carbon generation such as coal with CCS and renewable technologies, though potentially some gas power plants could still be part of the electricity mix[255]. Whilst lower than when all options are available, a degree of diversity could therefore remain a feature of the electricity system. As we outlined in Chapter 3 of the nuclear consultation document and in this White Paper, a mix of technologies with diverse characteristics is

---

254 *Distributed Energy – Initial Proposals for More Flexible Market and Licensing Arrangements* http://www. ofgem.gov.uk/Networks/ElecDist/Policy/DistGen/Documents1/DE%20con%20doc%20-%20complete%20 draft%20v3%20141207.pdf.

255 As shown in Box A3 above, when nuclear power is not available, other sectors have to make more efforts as abatement options are relatively cheaper than further reducing emissions from the electricity sector. We will need still some conventional gas plants to ensure sufficient flexibility in meeting demand in the electricity sector.

key to the security of the electricity system, as it avoids being overly dependent on any one technology so that the whole system is less exposed to any technology specific risk. The actual level of security of supply will then be determined by the reliability of those technologies in the energy mix.

A31 Coal and some renewable energy technologies rely on relatively reliable supply chains. Coal is the most abundant fossil fuel in terms of proven remaining reserves. The IEA estimates that at current rates of consumption, coal reserves would last for more than 150 years. Coal is also traded in global markets and responsive to changes in demand and supply over reasonable time horizons. In addition, it is comparatively easy to transport and store. Some biomass used in the electricity sector is likely to come from domestic sources, though as consumption increases we will have to import more and more. The market for biomass fuels is still relatively underdeveloped, though it is expected to grow considerably over the next decade. Renewable technologies such as wind, marine and solar are domestic by nature, and therefore do not rely on a long supply chain for fuel, even though they do rely on complex supply chains for the resources and skills needed to build them. In comparison nuclear power fuel supply is a stable and mature industry. Based on the levels of global nuclear generation in 2004, the known available reserves of uranium would last for the next 85 years. Uranium imports also come from a range of countries that are not necessarily the same as those that supply other energy sources. Uranium is currently mined in 19 different countries and resources of economic interest have been identified in at least 25 other countries.

A32 There are, however, potential risks of relying on a mix of electricity generation technologies with less diverse characteristics. Different technologies bring different characteristics and therefore play different roles in the energy market. As outlined in Chapter 3 of this White Paper, the characteristics of nuclear power enhance security of supply by (i) providing reliable low-carbon baseload electricity generation, (ii) increasing the diversity of supply sources, and (iii) reducing the vulnerability to volatility in fossil fuel prices. Excluding a technology with a different set of characteristics like nuclear magnifies the risks and uncertainties around the remaining set of technologies so that the overall electricity generation system is less flexible. This is especially true where these technologies are either yet unproven or not deployed on a large-scale, as is the case with CCS and some renewable technologies[256].

A33 Some technologies have potential limitations which need to be taken into account as the electricity system becomes more reliant on them.

256 We recognise that concerns were raised in the responses to the consultation about safety risks related to accidents or the threat of terrorist attacks. Some argued that these types of risks made nuclear power fundamentally different from other types of energy. The models we used in our analysis did not attempt to monetise all costs and benefits, for example a monetary value associated with potential accidents was not estimated. Evidence suggests that the likelihood of such accidents is negligible, particularly in the UK context. Though accident risk should not be dismissed, the assumption is that this can be managed through design of regulatory and corporate governance arrangements for the nuclear industry. This assumption is similar to the position of the Sustainable Development Commission, see *The role of nuclear power in a low carbon economy*, SDC position paper, March 2006. See Nuclear Cost Benefit Analysis (available at: http://www.berr.gov.uk/files/file39525.pdf) for more details.

Many forms of renewable generation are intermittent, and depend on external forces that are not always available (for example tides or wind). Some types of renewables, such as wind, are variable but their output is not very predictable while others, such as tidal power, are variable though predictable. Their contribution to the UK's electricity system will therefore be different[257]. Nevertheless, the system operator will have to take into account the intermittency of renewable energy when calculating how much reliable generation will need to be available at peak times. The system operator will also need to take into consideration that such generation may be limited in its ability to respond to short-term market signals. As a result, back-up generation would be required to maintain an adequate supply of electricity at all times. This would increase the amount of generation capacity required and the investment needed to strengthen networks to accommodate such new capacity.

A34   Without appropriate back up, the greater the percentage of intermittent renewables in the generating mix, the greater the risk to security of supply in the absence of increased demand-side response and/or expensive bulk storage of electricity. In particular, very high proportions of wind or marine generation could create difficulties for system operation and load balancing. The generation costs could therefore be higher because of the greater costs of system balancing resulting from the intermittent nature of renewable energy, and the costs of retaining or building thermal plant to maintain the same level of security of supply where renewables make up a very large proportion of the mix.

A35   If the low-carbon benefits of adding renewable capacity are not to be eroded, then this back-up capacity also needs to be low-carbon. This back-up could be biomass or coal with CCS, though it is likely that cheaper high-carbon options such as gas-fired generation would be used[258].

A36   There are also uncertainties over the speed with which some of the new and less developed technologies such as wave and tidal power but also CCS will develop. These uncertainties affect estimates of the likely timing, cost and feasibility of their deployment on a much larger scale. Important considerations of local acceptability, e.g. associated with alternative land use or wider environmental impacts, could also significantly hinder the deployment of such technologies and limit the number of available sites even further[259].

A37   Similarly, CCS represents a significant technological challenge. No commercial scale power station using CCS technology has yet been demonstrated anywhere in the world, although all the key elements of the individual stages of the process have been demonstrated. There

257 Tidal generation is predictable and can provide consistent electricity twice a day, similar to 'baseload' plant. It cannot, however, provide flexibility at times of peak demand.
258 According to the Redpoint model in fact, in the absence of nuclear power, carbon emissions from the generation sector could increase by up to 29Mt $CO_2$ in 2030.
259 A study of the potential for tidal power in the UK carried out by the Sustainable Development Commission suggests that around 90% of the UK's (practical) tidal range resource exists within the Severn Estuary and that a Severn Barrage could provide up to 5% of our electricity demand. Opposition to such a project would therefore limit considerably the potential capacity from tidal resources in the UK.

**167**

are technical uncertainties related to the construction of a system to transport and store carbon dioxide. There are also practical uncertainties related to applying the technology to electricity generation and to ensuring and monitoring the long-term integrity of the storage site, after injection has ceased. Based on the current status of CCS, there is therefore a high risk attached to placing too much future reliance on the ability of CCS to reduce carbon emissions, whilst retaining secure electricity supplies.

A38　An increase in the amount of Distributed Generation (DG) on the system would also represent a major challenge as it would require significant changes to how the system currently operates. DG potentially adds to the complexity of the role undertaken by the system operator (National Grid) in ensuring that electricity supply and demand remains in balance minute to minute[260]. Higher penetration of DG would significantly increase the number of generators bringing electricity onto the grid, requiring improved coordination between the system operator (National Grid), Distribution Network Operators and suppliers.

A39　Increasing DG capacity that is effectively invisible to the system operator (National Grid), therefore, could increase the level of uncertainty that has to be managed to ensure supply security. However, National Grid already copes with the vast numbers of customers whose demand is continually fluctuating throughout the day. A wider range of generators is essentially little different. Equally, the provision of electricity by a much wider range of producers reduces the importance of any one generator, and potentially makes the system much more robust to equipment failure and other temporary outages.

A40　Overall, therefore, without the option of nuclear power, we will be reliant on a less diverse mix of technologies to insure us against the future developments that could undermine security of supply, for example higher fossil fuel prices or disruption in the fuel supply chain. Some of these technologies, such as CCS and renewables, may not be deployable on a large scale to timely meet demand when needed. This would expose us, in some scenarios, to a higher risk of interruptions to electricity supply or to higher costs for delivering a given level of security of supply.

## Costs

A41　Achieving our long-term targets to reduce carbon emissions will require a considerable change in our energy and electricity system. We will have to make these changes whilst maintaining secure and reliable energy supplies. This will require significant new investment, both in the development of new low-carbon technologies and in the deployment of new electricity generation capacity, based on existing and new technologies. The costs for the economy will therefore be considerable. On the other hand, there is scope to improve the

---

260 See Ofgem's website for details on transmission and distribution networks: http://www.ofgem.gov.uk/ Pages/OfgemHome.aspx.

productivity of the economy, both by improved energy efficiency but also from the use of new and more advanced technologies.

A42    Some of the costs to be incurred will not necessarily be financial, but will arise from the changes we will need to make to our own lifestyle and from some of the opportunities we will have to forego to achieve our objective to move to a low-carbon economy. Such costs will occur regardless of which technologies become available, but they are likely to be even higher if we exclude some technology options, as this then limits our opportunities to balance the different types of risks.

A43    In particular, modelling for the Energy White Paper shows the range of economic costs that the UK will incur to achieve its 2050 targets for reductions in carbon emissions, under different scenarios, including those scenarios where new nuclear is not available. The Markal model estimates that delivering the 60% goal for $CO_2$ reduction in the scenario where we exclude new nuclear power stations, and all other technologies become available and are successfully deployed by 2050 at the cost assumed in the modelling, would by 2050 imply an additional annual cost to the UK economy of £1 billion compared to a scenario where nuclear power is available.

A44    If, in addition, the application of CCS on power generation were not to prove feasible, the model estimates that the cost in 2050 of achieving the 60% goal is likely to be at least an additional £5 billion per annum, compared to a scenario where all options were available[261]. As part of the impact assessment for the Climate Change Bill further analysis was conducted on achieving targets of 70% and 80% $CO_2$ emissions reduction by 2050. The costs of not allowing nuclear as an option would then rise to £3 to £5 billion per annum in the case of a 70% reduction or £3 to £11 billion per annum in the case of an 80% reduction target by 2050[262].

A45    However, as we mentioned above, any analysis, and particularly an analysis that makes use of models, cannot fully capture the full financial and social implications, especially in scenarios that exclude a particular technology[263]. This is particularly important because so long as their characteristics are distinct, widening the range of low-carbon electricity generation technologies will mean we are better able to meet our carbon and energy security objectives under the widest range of future circumstances. As their characteristics are distinct, widening the range of low-carbon electricity generation technologies will mean we are better able to meet our carbon and energy security objectives under the widest range of future circumstances. Moreover, as we reduce the number of options available to meet our objectives, the range of uncertainties around the remaining technologies becomes much more significant. In these scenarios we lose flexibility in the system. In other words, if a technology fails we have fewer alternatives. The Markal

261 By 2050, the total cost of achieving the 60% goal would be £21 billion (compared to no action to reduce emissions) with all options available, £22 billion when nuclear is not available and £26 billion absent both CCS and nuclear.

262 Final impact assessment for the Climate Change Bill is available at: http://www.defra.gov.uk/environment/ climatechange/uk/legislation/pdf/cc-impact-assessment-final.pdf.

263 See Box A1 for an explanation of the limitations of the modelling work.

**169**

model, in addition, does not take account of the security of supply considerations of relying on a set of technologies with less diverse characteristics. For example it does not consider our exposure to a greater risk of technological failure or exposure to the risk of a fuel supply interruption, for example in gas supply. The Markal model is also limited in its ability to capture the costs of maintaining the reliability of the electricity system as the share of intermittent generation in the mix increases[264].

A46 Nor does the model capture some of the risks inherent in the modelling assumptions. For example, it cannot model the risk that the costs of alternative low-carbon technologies do not fall as much as projected or the risk that we fail to see the behavioural change required to deliver the improvements in energy efficiency necessary to meet our 2050 goal. For this reason, we believe the cost estimates that the Markal model provides are likely to be at the lower end of estimates of the expected costs. The model is, however, very useful in illustrating the broad economic and structural impact of achieving our long-term targets for carbon emissions.

## Conclusion

A47 It is very difficult to predict how energy supply and demand and the electricity generation mix will develop over the very long-term. The factors which contribute to this uncertainty include: the cost and availability of fossil fuels, the cost and availability of emerging low-carbon technologies, and growth in energy demand.

A48 Our analysis indicates that narrowing the range of available low-carbon technologies make it more difficult for us to meet our energy policy goals under all circumstances. The Government believes that by having a diverse range of options, the UK will be better placed to deal with the range of possible futures that could unfold.

A49 The analysis shows that to achieve our target to reduce $CO_2$ emissions at least cost:
- All sectors of the economy will need to contribute in the effort to reduce $CO_2$ emissions
- The electricity sector will need, over the long-term, to considerably (if not fully) decarbonise, since reducing emissions from the generation sector is relatively less expensive than reducing carbon emissions in other sectors (for example transport). Low-carbon technologies will have to replace virtually all existing generation capacity by 2050
- Large changes will be needed in the electricity system in terms of the scale of new capacity needed: the EU 2020 Renewables targets will mean rapid deployment of renewable technologies in the medium-term while learning how to maintain security of supply with large penetrations of wind and other intermittent renewable

264 For an analysis of the potential costs from intermittency, see a report from UKERC on the estimated costs of integrating intermittent generation into the electricity system – available at: http://www.ukerc.ac.uk/ResearchProgrammes/TechnologyandPolicyAssessment/TPAIntermittencyReport.aspx.

technologies, most likely through considerable investment in backup capacity

- The overall challenges of delivering secure electricity supplies while making the transition to a low-carbon economy will be magnified over the long-term in the absence of a dependable low-carbon technology such as nuclear power
- This will be particularly significant should safe and reliable CCS for power generation not be proven or deployed at scale at reasonable costs
- Without nuclear power as an option, it would take a greater effort to reduce emissions through more costly options both within and outside of the electricity generation sector, and we would have to rely on generation technologies which together have a less diverse set of characteristics. This would expose the UK to greater risks of supply interruption and high prices, because our electricity system would not have access to a relatively low-cost, dependable low-carbon source of generation
- Our analysis suggests that excluding nuclear power as an option would therefore increase the risks and make it more expensive to meet our goal to reduce carbon dioxide emissions and to maintain secure energy supplies.

A50    In this context, Government's view is that our energy policy should promote, and be open to, all the technology options that are consistent with our goals for reducing carbon emissions and for achieving security of supply.

# Justification and Strategic Siting Assessment processes

B1    Alongside the Government's consultation on the future of nuclear power, the Government also consulted on proposed Justification and Strategic Siting Assessment processes[265]. This Annex provides an assessment of the key arguments submitted to the technical consultation and the Government's response.

## Justification

### Overview

B2    The concept of Justification is based on the internationally accepted principle of radiological protection that no practice involving exposure to ionising radiation should be adopted unless it produces sufficient benefits to offset the health detriment it may cause. This principle has been incorporated into European Community law by article 6(1) and (2) of Directive 96/29/Euratom. These articles were implemented in the UK by the Justification of Practices Involving Ionising Radiation Regulations 2004[266]. The process the Government has put in place for assessing applications for Justification is based around the Justification Regulations.

B3    A number of the responses to the Government's technical consultation suggested changes to the proposed Justification process that would require applicants to provide information that went beyond the scope of Justification as defined by the regulations or would be inappropriate for a high-level assessment. For example, some respondents made comments on how Justification would apply to specific sites. As Justification is a generic process, and site assessments are dealt with in detail at other parts of the regulatory process, it would not be appropriate to consider them here. Likewise, the Justification Regulations do not require a comparison with other forms of electricity generation and it would therefore be inappropriate to require applicants to do so. In such situations, the Government has not been able to accept the suggestions put forward.

B4    An assessment of the key arguments submitted to the technical consultation and the Government's response is as follows.

---

265 The Future of Nuclear Power, *Consultations on the proposed processes for Justification and Strategic Siting Assessment*, URN 07/972, May 2007.
266 Justification of Practices Involving Ionising Radiation Regulation 2004 (S.I.2004/1769).

## 1a. Are Government plans to structure the proposed Justification process by making a time-limited "call for applications" helpful?

### Key arguments and issues presented in responses

B5    There was broad support from those responding to the technical consultation for the Government's proposed Justification process. Many felt that the time limited call for applications would help create an impetus to the process for building new nuclear power stations and help focus resources and some suggested that the time limit should be no longer than two months. Some noted that the time limited call for applications did not preclude applications at any other point, however some felt that applications submitted during the window should be given priority.

B6    Some respondents felt that the time limited call for applications would encourage a number of applications to be submitted at the same time, which would allow the Justifying Authority to consider them together where appropriate. In this context, some respondents felt that it should be possible for a single application to be made covering a range of reactor designs, providing the health detriment and benefits were similar. This point is picked up in greater detail under Question 1b.

### Government response

B7    The Government can confirm that it will be issuing a time limited call for applications in February/March 2008. While this will not preclude applications being submitted at any other time, applications submitted during the call for applications will be processed as a priority.

## 1b. Is the proposed application, assessment and decision-making process clear, appropriate and proportionate? If not, how can it be improved?

### Key arguments and issues presented in responses

B8    A number of respondents felt that the process was clear, appropriate and proportionate, although some felt that the process was unnecessarily elaborate compared to the way in which other EU Member States handle Justification where there was no separate Justification process. A number felt that this would result in any decision taking longer than necessary. However, some felt that the process and timeframes were necessary given the importance of the Justification Decision.

B9    A number of respondents suggested that any further guidance on the proposed Justification process should make clear that the Justifying Authority will need to consider whether any new class or type of practice constitutes an existing practice which is already justified. Some felt that the Justification process should be as open and transparent as possible and that there should be some form of public engagement.

B10 Some respondents thought that any additional guidance should provide a clear definition of "health detriment" and on what constitutes a "practice". On the latter point, it was felt that the Justification Regulations talked of a "new type or class of practice" rather than a specific design within a particular class or type of practice, and suggested that this could be made clearer. They felt that this was inconsistent with the Government's technical consultation which referred to "nuclear power station technologies".

B11 A number of respondents felt that it should be possible for a single application to be made covering a range of reactor designs. This would involve establishing a technology envelope within a class or type of practice, which would be defined on the basis of similar potential health detriments and benefits.

## Government response

B12 The process that the Government is putting in place to assess Justification applications for new nuclear power stations is based on the existing Justification Regulations[267]. The Government believes that this process is fair, transparent and robust.

B13 The Government will produce guidance on the process for considering Justification applications in relation to new nuclear power stations, which will sit alongside Defra's Justification Guidelines[268]. This guidance will provide detail on the process for submitting and assessing Justification applications relating to new nuclear power stations and will provide clarity as appropriate.

## 1c. Is the indicative list of information, described in Appendix A[269], appropriate for applicants to be able to make an application?

## Key arguments and issues presented in responses

B14 A number of respondents felt that the indicative list of information was comprehensive, although some asked for clarity on the required depth and breadth of any application. For example, would it need to cover aspects of the fuel cycle that were already justified; that occurred outside the UK; or were common with other industrial activities? Some felt that the risks associated with terrorism needed to be considered as part of the Justification process.

B15 Some respondents suggested it would be helpful for any guidance to include a list of all the radiological health detriments that may

267 Justification of Practices Involving Ionising Radiation Regulation 2004 (S.I. 2004/1769).
268 www.defra.gov.uk/environment/radioactivity/government/legislation/justification.htm
269 Appendix A to The Future of Nuclear Power, *Consultations on the proposed processes for Justification and Strategic Siting Assessment*, URN 07/972, May 2007.

arise as well as a full list of all the potential environmental benefits and detriments to be considered. Some asked for clarity on whether detailed information on 'secondary' activities such as fuel manufacture and transport, which the respondent believed were existing practices, would be required and some felt it would be helpful to have clarity on which non-health detriments applicants should consider. Some respondents suggested the need for information about the lifecycle carbon footprint of proposed plants and a full cost/benefit calculation around each proposal.

B16 A number of respondents believed that a full analysis of potential radiological health detriments was necessary, while decisions on which benefits to incorporate was a matter for the applicant.

## Government response

B17 The specific guidance the Government is producing will provide additional detail on the process as it applies to new nuclear power stations.

B18 While it is for any applicant to include what information they feel is necessary and relevant to their application, the Government will, where appropriate, provide clarity on what information must and should be provided. However, this will provide guidance only and the Justifying Authority has the power to require additional information to be provided with respect to any application.

*1d. The Government is planning, where possible, to consider concurrent applications for Justification (relating to new nuclear power station technologies) through a single Justification assessment process.*
*Is the Government's proposal appropriate?*

## Key arguments and issues presented in responses

B19 A number of respondents felt that the approach was appropriate and would enable reactor designs to be assessed more quickly, although some felt that considering concurrent applications would add complexity and could delay the process.

B20 Some respondents felt that an application defined by a broad envelope of benefits and health detriments, within which a number of technologies could be shown to fit, would be suitable for a single Justification assessment. However, it was noted that any designs considered under a single Justification assessment needed to have similar health detriments and benefits.

B21 It was suggested that this approach was consistent with the Justification Regulations, which talked of a "new type or class of practice" rather than a specific design within a particular class or type of practice. It was suggested that this was inconsistent with the

Government's technical consultation which referred to "nuclear power station technologies".

## Government response

B22 The Government believes that considering concurrent applications for Justification through a single assessment will allow both applicants and the Justifying Authority to focus resources. This will enable the Justifying Authority to assess a number of designs with similar health detriments and benefits together, which may reduce the number of individual applications and therefore the number of assessments.

B23 The Government confirms that it should be possible to assess an application defined by a broad envelope of benefits and health detriments, within which a number of designs could be shown to fit, as a single Justification assessment. The Government can also confirm that this would only be possible if the designs had similar health detriments and benefits. However, the Government will still need to consider whether an application can be treated as relating to a single class or type of practice when it receives the application. This approach is consistent with the Justification Regulations.

## 1e. Are there any other ways in which the draft Justification process can be improved? If so, we welcome your suggestions.

### Key arguments and issues presented in responses

B24 A number of respondents submitted views on how the proposed process could be improved. For example, some felt that it would be helpful to set out a specific timetable for the Justification process and decision, including any plans for public engagement.

B25 Some suggested publishing a list of nuclear technologies that would not be considered and processing designs that had already been Justified by other Member States more rapidly.

### Government response

B26 The guidance the Government is producing will provide additional detail on the process for submitting and assessing Justification Applications along with an indicative timeframe and any plans for public engagement. The Justification Regulations do not cater for the fast tracking of designs justified outside the UK.

# A combined Strategic Siting Assessment and Strategic Environmental Assessment process

## Overview

B27 The consultation document on the Strategic Siting Assessment (SSA) set out a proposed process for determining the suitability of potential sites for new nuclear electricity generation and identifying siting criteria. It also set out proposals for conducting a Strategic Environmental Assessment under the Strategic Environmental Assessment Directive (SEA)[270]. The UK's own implementing regulations require such factors to be taken into account in developing plans or programmes which will have consequences for the environment. The consultation document on the SSA presented a site selection approach which incorporated the SEA into the Strategic Siting Assessment. The results of the SSA would inform a subsequent Governmental policy statement on siting for new nuclear power stations, as part of a potential National Policy Statement (NPS) on new nuclear power stations.

B28 Respondents to the consultation provided a range of comments on the proposed SSA process. Some agreed that the SSA process was logical and robust, and that the approach incorporating the SEA was a reasonable one. Others commented that the process could be addressed by building on existing sites or that it should be evident early on in the process which areas of the country are suitable or those which can be ruled out, which led to questioning whether there was a need for a detailed process as set out in the consultation document, and whether there was scope for compressing stages 1 and 2. Others commented on the need for greater clarity on whether the SSA would provide a list of existing sites or localities, or a particular grid reference. There were, however, a number of themes which emerged from the responses and these are set out below in response to each of the questions in the Technical Consultation Document.

## 2a. Is the proposed approach to the Strategic Siting Assessment a logical approach to identifying suitable sites? If not, how could it be improved?

### Key arguments and issues presented in the responses

B29 Some respondents commented that the final statement for the SSA should not restrict the eventual number of sites or be exhaustive. The reasons for these include:
- enabling a workable market in sites
- some sites which might be successful in the SSA process may be found to be unsuitable at site-specific level when more detailed work is carried out, thereby reducing the number for developers to choose from

---

270 Directive 2001/42/EC of 27 June 2001 on the assessment of the effects of certain plans and programmes on the environment (O.J. L197, 21.7.2001, p.30).

- enabling other suitable sites to come forward at some point in the future
- that it impacts on the ability to deliver any substantial programme.

B30 Although no specific question was asked, a number of respondents provided detailed comments on the criteria. These comments included:
- the need for the SSA to limit the number of exclusionary criteria, as it was felt that there were, perhaps, only one or two criteria such as population density that were truly exclusionary at the national level, and not those for which there might be technical solutions
- those criteria which have an economic effect or could be mitigated against should be treated as discretionary criteria
- using the opportunity of the SSA to define appropriate demographic criteria taking account of developments in reactor design and UK and international experience, and that such criteria should be used as the basis for siting policy
- that it would be appropriate for criteria on, for example, flood risk management, effect of climate change, and public acceptability
- consideration should be given to issues of staffing such a facility.

B31 A number of comments were related to the nominations process mentioned in the SSA process. In general, the comments welcomed the nominations process. Specific comments were made on the need to provide an early indication of guidelines on who can make a nomination and the information required to support a site. It was pointed out that the information for the nomination phase should be limited to publicly available information and that a nomination must not require on-site studies.

B32 There were some comments on the interaction with the planning process. Some respondents said that the SSA seemed to be appropriate so long as the normal planning processes were followed for developments. Others supported the intention to include the outcome of the SSA as a material consideration in a NPS which would provide the framework for consent for the independent Infrastructure Planning Commission. Others felt that the requirements for a NPS should be included in the SSA, such as the need for local engagement at the sites likely to be affected, and that such consultation should happen at Stage 3 of the SSA, which should not affect the timescales for the SSA overall. Other comments related to the validity of the NPS, and questioned whether there would be enough sites on the final list for the NPS to be sufficiently durable to enable replacement of nuclear capacity.

B33 There were some general comments and some concerns about the timescales proposed for the SSA and SEA in the Technical Consultation Document. Some respondents felt that the timescales as set out were realistic and consistent with ensuring effective consultation takes place, but that it was essential that the SSA sticks to the 18 months-2 years timescale. Others felt that the process was over-elaborate and could lead to unnecessary delays.

B34 A number of respondents commented on the geographical scope of the SSA/SEA. Some said that these should be UK wide, whilst others said they should in particular cover England, Scotland and Wales. Some respondents expressed concern that the approach of the Devolved Administrations could delay development in those areas.

## Government response

B35 We acknowledge the concerns that people have raised on the SSA. In the main, it will be appropriate to address these concerns as we take the SSA forward as they relate to the implementation of the process, in particular in the development of the criteria and the nominations process. However, it is worth noting here that in response to concerns raised with regard to the planning system and the need to ensure alignment with the planning reforms, Government has considered the scope for bringing the requirements for the NPS more closely into Stage 3 of the SSA. The Planning Bill reforms propose consultation with those local communities likely to be affected by the proposals and a requirement for parliamentary scrutiny. In taking forward the NPS, we will build these elements into it as necessary. Also in response to the concerns raised regarding the timescale, Government will endeavour to limit the slippage which could potentially arise. We have set out the process we will take forward for the SSA and SEA in Section 3 of this White Paper.

## 2b. Does the proposed incorporation of the SEA into the SSA represent a reasonable and robust approach to assessing environmental issues that would be raised by the construction and operation of new nuclear power stations?

### Key arguments and issues presented in the responses

B36 Respondents made a range of points on the incorporation of the SEA into the SSA. Some respondents felt that the incorporation of SEA is important to ensure that the SSA is as comprehensive and legally robust as possible. Some also commented that the process looks robust and reasonable from the level of detail provided but that details on how the SEA would apply need to be clarified. Some respondents stressed the importance of the SEA as an integral part of the SSA which would help to minimise the possibility of a further SEA after completion of the SSA, and before site-specific planning proposal could be considered. Some respondents were concerned to ensure that the timing of the different stages are integrated properly so that the process is capable of delivering a list of sites that meet the criteria, whilst ensuring that environmental and other effects such as health and socio-economic impacts are properly assessed.

B37 A number of respondents expressed concerns about the iterative nature of the SEA, and the scope for delay and duplication. Some respondents thought that the SEA may extend the overall timescale beyond that

**179**

outlined in the consultation which would produce uncertainty about the eventual availability of sites. Others, however, felt that the SEA could be completed thoroughly in the timescale available. Overall, respondents felt that Government would need to ensure that the iterative nature of the SEA process is carefully managed so as not to increase the timescales and add delay to the overall timetable.

B38   There was also some concern around the potential for duplication of issues covered by the SEA and the Environmental Impact Assessment (EIA) which developers need to complete as part of the planning process. Some respondents felt that the SEA should be applied at the locality rather than site level so that the assessment remains strategic and does not overlap with the EIA.

B39   There were some comments on the interaction with other policies. Some respondents felt that the SEA for the SSA would need to be able to support the NPS, and that this should be covered in any scoping document for the SEA.

## Government response

B40   We acknowledge the concerns raised in relation to the SEA. As with the SSA, we will consider these in taking forward the SEA, for example ensuring that there is a minimal overlap between the SEA and the EIA, and ensuring that the iterative nature of the SEA does not lead to unnecessary duplication and delay. We also take on board the concerns expressed on the need to ensure that any requirement for a SEA for the NPS is covered by the SEA from the SSA. We will consider the best way to link these processes as we move forward with the SEA.

B41   The proposals at Section 3 of this White Paper set out how the SSA and SEA will be taken forward as an integrated approach.

ANNEX C

# Regulatory and advisory structure for nuclear power

C1    This Annex explains aspects of the existing and future regulatory and advisory committee structure for nuclear power.

## Oversight of nuclear power stations in the UK

C2    Government recognises that the way in which any new nuclear power stations might be consented, built, operated and decommissioned is an area of particular concern to many people. The purpose of this Annex is to outline the protections currently in place, and which would apply to any new nuclear power stations, to ensure that these processes are carried out safely and effectively. We also set out the terms of reference of the re-constituted Committee on Radioactive Waste Management, and describe the role of a new Nuclear Liabilities Financing Assurance Board.

C3    In Great Britain, the main regulatory bodies are the Nuclear Installations Inspectorate (NII), a division of the Health and Safety Executive, the Environment Agency in England and Wales and the Scottish Environment Protection Agency in Scotland.

C4    These agencies regulate radioactive discharges from nuclear power stations and have responsibilities (see below) for ensuring that workers, the general public and the environment are protected against exposure to radioactivity. In Northern Ireland (NI) the relevant authorities would be the Secretary of State, HSENI and the Department of the Environment.

C5    Nuclear security is the responsibility of the Office for Civil Nuclear Security (OCNS) which has been part of the HSE since April 2007. It places strict obligations on operators and requires site security plans to be regularly reviewed. For any new build, the OCNS will ensure that security measures are included in plans for the construction of any new nuclear power stations from the outset. Doing so will avoid the need for retrofitting security measures once construction is underway and will enable regulators to make an early judgement with regard to establishing the most appropriate measures at any construction site should approval be given[271].

C6    New powers to be introduced in the Energy Bill will put into place a framework to ensure that operators of any new nuclear power stations

271 http://www.hse.gov.uk/nuclear/ocns/ocns0607.pdf

pay their full decommissioning costs and their full share of waste management and disposal costs.

C7    This framework requires operators to provide and have approved a programme, outlining how they intend to manage waste and decommissioning, along with detailed costing of these plans and proposals for how these costs will be financed.

C8    In recognition of concerns raised in the consultation, we intend to create a new independent advisory body, the Nuclear Liabilities Financing Assurance Board (NLFAB). This new board will provide independent scrutiny and advice on the suitability of the decommissioning programmes submitted by operators of nuclear power stations.

C9    The NLFAB will advise the Secretary of State for Business, Enterprise and Regulatory Reform on the financial arrangements operators plan to put in place to cover waste management and decommissioning. The NLFAB will also advise the Secretary of State on the regular reviews and ongoing scrutiny of funding arrangements, once new nuclear power stations are operational.

C10   The Board is expected to consist of experts from relevant fields such as current or former fund managers, pension trustees, actuaries and nuclear engineers. The board members will be appointed by the Secretary of State.

C11   The NLFAB will have a tightly defined, solely advisory role. Its work will focus on ensuring that the operators of new nuclear power stations put in place robust financial arrangements for clean up.

## Safety regulation

C12   The Health and Safety Executive has statutory responsibility for ensuring that there is an adequate framework for the regulation of safety at nuclear sites in the UK. This responsibility covers the licensing and day-to-day regulation of nuclear sites and the regulation of work-related health and safety generally.

C13   The legal framework requires nuclear operators to demonstrate to the satisfaction of the HSE's Nuclear Installations Inspectorate (NII) the safety of activities at nuclear sites and that they are complying with the strict conditions of their nuclear site licence, and other relevant safety legislation.

C14   Licensing applies throughout the lifetime of a nuclear installation from design and construction to eventual completion of decommissioning and clean-up. Licence conditions cover all the arrangements for managing safety, including the production of adequate safety cases for all operations, the appointment of competent personnel, staff training and supervision, handling and storage of nuclear material, control of

**182**

organisational change, response to accidents and emergency planning arrangements.

C15   NII inspects nuclear sites and scrutinises operators' safety cases to ensure that the evidence they present is robust. Safety cases are frequently required before NII will consent to the start of certain operations, such as restarting a reactor after major maintenance. In addition, licensees must review and re-assess the safety of their plants periodically and systematically, generally every ten years. HSE's reports on licensees' Periodic Safety Reviews (PSRs) are usually published.

## Security regulation

C16   The Office for Civil Nuclear Security (OCNS) is the security regulator for the UK's civil nuclear industry. It is part of the Health and Safety Executive (HSE) and is responsible for approving security arrangements within the industry and enforcing compliance. OCNS conducts its regulatory activities on behalf of the Secretary of State for Business, Enterprise and Regulatory Reform and under the authority of the Nuclear Industries Security Regulations 2003.

C17   OCNS also undertakes vetting of nuclear industry personnel with access to sensitive nuclear material or information. It works closely with BERR policy officials, other Government departments and with overseas counterparts.

C18   In the UK, civil nuclear operators must have site security plans dealing with the security arrangements for the protection of nuclear sites and radioactive material on such sites. The arrangements cover, for example, physical protection features such as fencing, CCTV and turnstile access, the roles of security guards and the CNC, the protection of proliferation-sensitive data and technologies and the trustworthiness of the individuals with access to them. Transporters of nuclear material also have to be approved by OCNS, acting for the Secretary of State, and approval of a transport plan is required before the transport of certain categories of nuclear material.

C19   OCNS may give directions to operators or carriers at any time, for instance in the light of a change in the threat level for the industry. This is aided by OCNS being an active member of the UK intelligence community.

C20   OCNS determines and keeps under review the numbers and tasking of the Civil Nuclear Constabulary's officer at licensed nuclear sites. CNC is an armed police force tasked with protection of nuclear material and nuclear sites.

C21   OCNS publishes an annual report on the HSE website[272] and publishes guidance for the industry. One key document covers the control of sensitive nuclear information, entitled "Finding a Balance".

272 www.hse.gov.uk

## Environmental regulation

C22 The Environment Agency and the Scottish Environment Protection Agency are the principal environmental regulators in England and Wales and in Scotland respectively. They have a number of regulatory roles in relation to nuclear sites. These include under the:
- Radioactive Substances Act 1993, regulation of all disposals, including discharges to air, water and land, of radioactive wastes off or on nuclear sites
- Water Resources Act 1991, regulation of abstraction from, and discharges to controlled waters (inland and marine surface waters, and groundwater)
- Pollution Prevention and Control Regulations 2000/ Pollution Prevention and Control (Scotland) Regulations 2000 (as amended), regulation of certain installations including, for example, combustion plant used as auxiliary boilers and emergency stand-by power supplies, and incinerators used to dispose of combustible waste
- Environmental Protection Act 1990 regulating disposals of waste by deposit on or into land, including excavation materials arising from construction; and acting as enforcing authority for the remediation of certain contaminated land which has been designated a "special site" in accordance with the Contaminated Land Regulations 2006.

C23 Additionally in England and Wales, local authorities or the Environment Agency usually take responsibility for flood defences. However, at nuclear sites operators take direct responsibility for their local flood defences as part of their safety obligations. To facilitate this, the Environment Agency usually makes agreements or other arrangements with site operators so that respective responsibilities are clear.

## Safeguards regulation

C24 Nuclear safeguards regulation aims to verify that States comply with their international obligations not to use nuclear materials (plutonium, uranium and thorium) for nuclear explosives purposes. Global recognition of the need for such verification is reflected in the requirements of the Treaty on the Non-Proliferation of Nuclear Weapons (NPT) for the application of safeguards by the International Atomic Energy Agency (IAEA). Also, the Treaty Establishing the European Atomic Energy Community (the Euratom Treaty) includes requirements for the application of safeguards by the European Commission.

C25 BERR is responsible for the UK Government input into the development of the international nuclear safeguards regimes. This aims to ensure that the IAEA safeguards regime is technically equipped to provide the assurances demanded of it by the international community (e.g. to develop and implement new safeguards strengthening measures), and also to ensure that nuclear non-proliferation policy properly reflects safeguards and verification-related considerations.

**184**

C26 Responsibility for overseeing compliance with the UK commitment to the international safeguards regimes belongs to the UK Safeguards Office at the Health and Safety Executive (HSE).

C27 The UK Safeguards Office (UKSO) is part of the Nuclear Directorate of the HSE and oversees the application of nuclear safeguards in the UK to ensure that the UK complies with its international safeguards obligations by:

- working with the UK nuclear industry and others with safeguards reporting requirements, and safeguards inspectors from the European Commission and the IAEA, to make sure that the safeguards measures applied are both effective and efficient
- ensuring that safeguards measures do not place unreasonable demands on, or result in unnecessary commercial disadvantage to the UK organisations involved
- helping to negotiate facility specific safeguards reporting and inspection arrangements with the European Commission and/or the IAEA
- assisting UK operators, especially those unfamiliar with the subject, in meeting safeguards requirements
- implementing the UK's Additional Protocol
- providing support to safeguards officials in BERR on safeguards policy issues that arise from the work of HSE (UKSO).

## Transportation of nuclear materials regulation

C28 The safety and security of nuclear material (including irradiated or spent nuclear fuel) is subject to rigorous regulation, which fully takes into account international obligations and commitments. These regulations meet the requirements of European Directives[273] for transport of radioactive materials as well as the International Atomic Energy Authority's Standard for the Safe Transport of Radioactive Material[274].

C29 The security for the transportation of nuclear material is regulated by the Office for Civil Nuclear Security (OCNS). OCNS is kept fully briefed about terrorist threat intelligence and in turn keeps security arrangements under review at all times. OCNS is satisfied with the thorough measures that have been taken to prevent the theft or sabotage of nuclear material in transit.

C30 The safety of nuclear transports (and security of less sensitive nuclear material) is regulated by the Department for Transport under The Carriage of Dangerous Goods and the Use of Transportable Pressure Equipment Regulations 2007.

---

273 Council Directive 94/55/EC of 21 November 1994 on the approximation of the laws of Member States with regards to the transport of dangerous goods by road.
Council Directive 96/49/EC of 23 July 1996 on the approximation of the laws of Member States with regards to the transport of dangerous goods by rail.
Council Directive 86/618/Euratom of 27 November 1989 on informing the general public about health protection measures to be applied and steps to be taken in the event of a radiological accident.
274 TS-R-1 (2005 Edition).

## Committee on Radioactive Waste Management (CoRWM)

C31    Following the announcements by UK Government and the Devolved Administrations, on 25 October 2007, a new CoRWM has been appointed under revised terms of reference. The Committee is jointly appointed by sponsoring Ministers from Defra, BERR and the Devolved Administrations.

C32    The role of the reconstituted Committee is to provide independent advice to Government on the long-term management, including storage and disposal, of radioactive waste. CoRWM's priority task will be to provide independent scrutiny on the Government's proposals, plans and programmes to deliver geological disposal as the long term management option for the UK's higher activity wastes.

C33    CoRWM is an Advisory Non-Departmental Public Body (NDPB).

C34    CoRWM shall consist of a Chair and up to fourteen members. Seats are not representative of organisation or sectoral interests and the skills and expertise which will need to be available to the Committee will vary depending on the programme of work.

C35    CoRWM will undertake its work in an open and consultative manner, engaging with stakeholders and publishing advice (and the underpinning evidence) that is meaningful to the non-expert, in an open and transparent way. The Committee will also undertake ongoing dialogue with Government, the Nuclear Decommissioning Authority (NDA), local authorities and stakeholders, and will liaise with appropriate advisory and regulatory bodies to provide an annual report of its work.

C36    CoRWM's advice, and the response of UK Government and relevant Devolved Administrations, will be made available to Parliament and Assemblies. Parliamentary and assembly committees will also have the opportunity to engage directly with CoRWM and may propose work for inclusion in the Committee's work programme to sponsoring Ministers.

Printed in the UK for The Stationery Office Limited
on behalf of the Controller of Her Majesty's Stationery Office
ID5667762  01/08

Printed on Paper containing 75% rcycled fibre content minimum.